9·26·75

A <u>PREFACE</u>
<u>TO</u>
ASTR<u>ONOMY</u>

by
Morris <u>Goran</u>

a TECHNOMIC® publication
TECHNOMIC Publishing Co., Inc.
265 W. State St., Westport, Conn. 06880

A PREFACE
TO
ASTRONOMY

© TECHNOMIC Publishing Co., Inc. 1975
265 W. State St., Westport, Conn. 06880

Printed in U.S.A.

Library of Congress Card No. 74-81581

Standard Book No. 0-87762-158-6

For
my Father
and the Memory
of my Mother

BOOKS BY MORRIS GORAN

Introduction to the Physical Sciences. Glencoe, III: Free Press, 1959.

Experimental Chemistry for Boys. New York: John F. Rider Publisher, 1961.

Experimental Biology for Boys. New York: John F. Rider Publisher, 1961.

Experimental Astronautics. Indianapolis: Howard W. Sams, 1967.

Experimental Earth Sciences. Indianapolis: Howard W. Sams, 1967.

The Core of Physical Science. Chicago: Cimarron Publishers, 1967.

Experimental Chemistry. London: Lutterworth, 1967.

The Story of Fritz Haber. Norman: University of Oklahoma Press, 1967.

Biologia Experimental. Barcelona: Ramon Sopena, S.A., 1967.

The Future of Science. Rochelle Park, N.J.: Spartan-Hayden, 1971.

Science and Anti-Science. Ann Arbor, Mich.: Ann Arbor Science Publishers, 1974.

PREFACE

One of the marks of the adjusted individual is the ability to transact with another person, group, one's self and with nature. Urban settings and modern mass education have emphasized the first three of the interactions. The last has not had as much guidance except for the environmental awareness fostered during the past few years.

The aim of this introduction to astronomy is to impart a broad, cosmic perspective for a better platform with which to assess our place in nature. In order to cater to a wide readership, mathematics has been kept to a minimum, both pure and applied aspects of the subject are treated, and human interest material is included.

The selected references at the end of each chapter are meant to be representative of the book literature on the subject. Many of them will lead the reader to pertinent periodicals such as *Sky and Telescope, Science News,* and *Scientific American*. Hopefully, some will pursue the numerous leads given in Richard Berendzen and David De Vorkin, "Resource Letter EMAA-1: Educational Materials in Astronomy and Astrophysics," *American Journal of Physics*, 41, June, 1973, p. 783–808.

TABLE OF CONTENTS

CHAPTER 1

THE EARLIEST VIEW OF THE WORLD

Long before science was an amateur activity, generations ago when man's chief pursuit was hunting or farming, his view of the heavens varied from one region to another. These conceptions can often be abstracted from literature, folk songs, artifacts and persistent myths.

One of the oldest systems of the world is given in the writings of Homer, the *Iliad* and the *Odyssey*. According to the poet, the ocean was original material from which came the earth. Probably disc-like in form, the earth floated on the immense ocean. Above was a solid hemispherical vault known as the firmament or sky, giving support for the heavenly bodies. The sun and stars were considered insignificantly small compared to the earth. Dark, subterranean caverns were in the submerged region of the earth while a mystic ocean stream flowed around its upper surface. Gods were above the firmament while the spirits of the dead were in bottom sections of the earth.

The starry vault made a daily rotation under the ocean. The planets also had their individual motions but the sun was held in such esteem that a special arrangement was for its conveyance. Apollo, the Sun-God "drove his fiery chariot daily over the celestial arch, carrying the glowing orb in its diurnal path, and the swift boats of the ocean stream conveyed it by night from the western exit back to its eastern gate."

The Elizabethan poet John Milton (1608–1674) probably had this legend in mind when he made Camus say:

> "The star that beds the shepherd fold
> Now the top of heaven doth hold;
> And the gilded car of Day
> His glowing axle doth allay
> In the deep Atlantic stream;
> And the slope sun his upward beam
> Shoots against the dusky pole,
> Pacing toward the other goal
> Of his chamber in the east."

Another fanciful theory of the early Greeks, albeit dealing with creation, can be found in the work of the poet Hesiod. He reported that chaos or disorder was the prime condition and from it were born the earth and love as well as darkness and night. Light and day came from the latter two while Uranos, the changeless sky, developed from the earth. All the major gods sprang from Uranos and the earth.

Cronos, or time, ruled all things. Cronos attempted to swallow his three sons at their birth but failed in the case of Zeus. He persuaded his father to disgorge himself of Poseidon and Hades and to divide his kingdom among his three children. Zeus, or Jupiter, inherited the kingdom of the sky, Poseidon, or Neptune, was heir to the oceans, and Hades, or Pluto, ruled the underworld. None of the gods ruled the earth's surface, common ground, but Jupiter and Pluto spent much of their time interfering with the affairs of men.

The early Babylonian views of the origin and nature of the world appears on the surface to be similar to that of the Greeks but differences are many. According to the ancient Babylonians, the mother goddess of dark chaos and the father ocean generated the gods of light. The mother goddess battled for supremacy with one of her progeny, son of the god of wisdom. The victor, Marduk formed heaven from one half of her and earth from the other. He also arranged the stars, sun and moon in an orderly pattern and the laws governing their motions.

The early Babylonians saw the universe as a double pyramid of square blocks, the upper tier of which formed the land of the living and the lower the home of the dead. Seven concentric spheres surrounded the double pyramid. Supreme beings lived in the upper sections between the spheres while the dome of fixed stars and the kingdom of the major gods was in seventh heaven. A treacherous ocean stream surged on the middle block that separated the upper and lower worlds.

Dante (1265–1321), in his *Divine Comedy* probably influenced the early ideas. In his book, the earth is a sphere concentric with the universe. Hell is a cone inside the earth reaching down to the very center. Ten solid spheres concentric with the earth are of increasing heavenliness: The lowest is that of the moon, then follow those of Mercury, Venus, the sun, Mars, Jupiter and Saturn. The eighth sphere is that of the fixed stars on firmament, the ninth or crystalline sphere in the primum mobile, and the tenth the empyrean, the dwelling of God. The outermost is perfect and motionless; the ninth sphere has a very high speed which is transmitted in diminishing degrees to all the inferior spheres.

Both Nordic and Chinese mythology are similar to the Babylonian in depicting anthropomorphic origins for the universe. The first had a giant from whose body the land was born and whose sweat created the sea. His skull was the source of the firmament, lighted by stars created by the gods who also harnessed the sun and the moon. The Chinese considered Tao as the "great original cause" who created a shaggy dwarf. Winds began from his breath and light from his eyes. He called into existence sun, moon and stars. When he died, the head became the mountains, his blood the rivers and his sweat the rains. Plants and trees came from his skin and hair. The human race arose from the insects on his body.

SELECTED REFERENCES

P. Lum, *The Stars in Our Heaven*, New York: Pantheon Books, 1952.
M. K. Munitz, Ed., *Theories of the Universe*, Glencoe, Illinois: Free Press, 1957.

CHAPTER 2

THE OBSERVABLE FACTS WITHOUT INSTRUMENTS

Men, women and children in all early cultures had some kind of familiarity with astronomical objects. They could approximate time, location and season; they could recognize familiar groups of stars; they could cite planets. Despite these talents they came, until recently, to erroneous views about the nature of the world. Since the relatively recent invention of electric lighting and our rapid urbanization, the average person has lost touch with astronomical sights. Nonetheless he may have a correct perspective of the nature of the universe.

Earliest man knew observable facts about the sky in his locality. Modern man can have the advantage of knowing what the sky is like in any region of the earth.

Within the forty eight states of continental United States, sun, moon and some planets are the astronomical objects in the day sky. (Some claim to be able to see a bright star if they are at the bottom of a deep well.) Sun, moon and planets travel in a regular fashion in the southern skies.

About September 21, the sun rises due east and sets due west. The next day its place of rise and set is slightly south of due east and west. Day by day the process of edging southward at dawn and dusk continues for about three months. About December 21, the sun reaches the farthest point south of east at its rise, and south of west at its set. During the three months thereafter, the sun edges back to the due east and west points, a little each day. For six months, from about September 21 to about March 21, the sun not only travels in the southern skies but is also rising and setting in the region.

About March 22, the sun at the break of day begins to appear slightly north of east, and as darkness starts to prevail disappears on the horizon slightly north of west. Each day thereafter the sun appears farther north of east and west. About June 21 the sun at dawn and dusk is farthest north of due east and west. During the three months after June 21, the sun edges back slowly, a little each day, to rising and setting due east and west. Thus for six months, from about March 21 to about September 21, the sun while traveling during the day in the southern skies is rising and setting in northern skies.

The yearly cycle is also related to height above the horizon of the sun at noon. It reaches its highest altitude about June 21 and its lowest about December 21.

Date	Place of Rise	Place of Set	Height at Noon
September 21	Due East	Due West	Average
December 21	Farthest South of East	Farthest South of West	Lowest
March 21	Due East	Due West	Average
June 21	Farthest North of East	Farthest North of West	Highest

The Sun During The Year

3

The sky region where the sun travels is called the zodiac. The name arises from the star groupings in the area. Aries, Taurus, Gemini, Cancer, Leo, Virgo, Libra, Scorpio, Sagittarius, Capricornus, Aquarius, and Pisces are largely animal configurations and the zodiac may stem from a term for a zoo. When the sun is out, it shines so brightly that the zodiacal constellations are not seen but at night, with the sun out of the way, the path of the sun through the stars is apparent.

For some, the moon is a night object although even careless observation reveals that about half of every month the moon is in day skies. Day or night, the moon is also seen in the zodiacal region.

Those planets visible to the unaided eye during the day are also in the same path of zodiacal constellations. If not on the horizon, planets are disk-like and brighter than stars; planets do not twinkle.

Stars twinkle except when seen on the horizon. The effect is due to a pin point of light being observed through the not-immobile earth's atmosphere.

Even with a polluted urban environment, the view of the stars and the night sky is great; it is one of grandeur when seen in the countryside. In clear air the innumerable dots of light seem too close and omnipotent.

Limited observation anywhere reveals several motions. The very speedy travel of falling stars is the fastest; they can be seen best after midnight. Another motion easily discernible is the east-to-west one of each object at the rate of $15°$ every hour. Then the sun, moon and planets, if watched over several days, can be observed shifting eastward against the background of stars. Finally, the planets, if followed for weeks, can be seen occasionally moving to the west.

Every object in the sky appears to move from east to west and makes a complete circuit in 24 hours. Called circumpolar rotation, or diurnal motion, it is the turn of all astronomical bodies about the north celestial pole. Closer examination with instruments reveals the movement is about a point, a short distance, about a degree, away from the star Polaris. Diurnal motion is apparent at night but the sun's movement during daylight hours is the same phenomenon.

Sun, moon and planets move continually eastward against the background of the so-called fixed stars at varying rates. The sun can be seen in the same position with respect to the stars at the same time only once a year. Since this circuit is $360°$ and the time is 365 and a fraction days, the sun's eastward drift is about a degree a day. Only experienced observers can make out such a change at dawn or dusk from day to day. Neophytes can appreciate the situation by comparing the star background at sunrise or sunset from month to month.

The moon's eastward drift is so large that even a nursery school child can see it. Simply compare the moon's sky position at the exact same time from one day to the next. The beginner can use a chimney, steeple or tall tree as a reference.

Each planet has a unique rate of eastward drift but like that of the sun's, a specious glance does not reveal the situation. Perhaps an interval of two weeks between sightings at the same time shows the effect clearly to those unfamiliar with the sky.

Only planets show an occasional westward, or retrograde motion, with the stars considered fixed in position with respect to each other. Each planet has a unique time for this apparently contrary exercise. Their eastward followed by westward travel is what caused some early observers to call them planets, meaning wanderers.

Those who first used the name lived in the Mediterranean Sea area. Here, as in continental United States, the zodiacal pathway is in the southern skies. With the exploration of the earth in later generations, the path was seen in other parts of the sky.

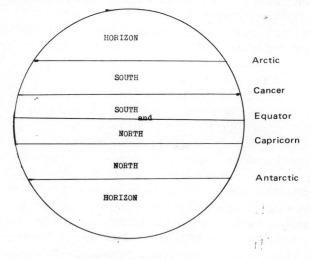

Location of Zodiacal Path

If an observer is between the tropic of Cancer, 23½° north of the equator and the Arctic Circle, 66½° north of the equator, the southern skies do have the zodiacal pathway. These are the regions where civilization has flourished. Here, houses seeking a great deal of sunshine face the windows to the south.

Those living between the tropic of Capricorn, 23½° south of the equator and the Antarctic Circle, 66½° south of the equator see the zodiacal path in their northern skies. Thus architects in Sao Paulo, Brazil and other cities in the region have requests for homes with much northern exposure.

In tropical areas, between the tropic of Cancer, 23½° north of the equator, and the tropic of Capricorn, 23½° south of the equator, the zodiacal path is in both their northern skies and southern skies. Twice a year, on March 21 and September 21, at the equator the sun is 90° above the horizon.

Not much is available on the earth above the Arctic Circle, 66½° north and the Antarctic Circle, 66½° south. In the latter area several countries have research stations. Both polar regions have six months of continual darkness followed by six months of continual daylight. The sun moves across the horizon at a low latitude and in both northern and southern skies.

SELECTED REFERENCES

Gerald S. Hawkins, *Splendor in the Sky*, New York: Harper and Row, 1961.

W. C. Saslaw and K. C. Jacobs, Eds., *The Emerging Universe*, Charlottesville: University of Virginia Press, 1972.

CHAPTER 3

THE EARTH

Evidence for the shape of the earth is often stated in a cursory manner. Either the way in which a ship is seen to disappear beyond the horizon or the return of a circumnavigating vessel is given as positive proof that the earth is spherical. However, both of the statements are more appropriately evidence for earth curvature, for an earth shape that could be like a banana or a doorknob.

Several other demonstrations indicate that the earth is curved and the nature of the curve cannot be determined by the evidence. For example, if three identical sticks are set into the ground along a straight line an equal distance but several yards apart, peering over their tops from one end reveals the middle stick to be higher. The phenomenon could not occur on a flat surface. A view from a rising balloon or airplane also shows a curved earth surface.

When modified, the above evidence can become suitable for sphericity. If a ship disappears at the horizon in the same fashion no matter the geographical region, then the curvature must be equal everywhere, and that can only mean a sphere. If a circumnavigating vessel operating at a uniform rate takes the same amount of time to cover great circles like the equator, then only a sphere is involved. If three identical separated sticks an equal amount in the ground are seen the same way wherever planted, a sphere is the only possible curve. If the airplane or balloon observer detects the identical curvature at equal heights at a large variety of earth stations, then a sphere is the only possible shape for the earth.

Photos of the earth taken from near the moon reveal a disk similar in shape to that of the planets. If the latter are considered to be spherical then the earth should be in the same category. This is not the best kind of argument to reach the conclusion, but a single, logic-tight evidence for earth sphericity is difficult to formulate. Galileo thought he had one citing the round outline when the earth's shadow covers the moon; he maintained that only a sphere can cast a circular shadow at all times. Yet such a shadow can be produced by a cylinder with a circular base.

A pragmatic approach to the problem is to claim that use of the concept of earth sphericity has not led to error but indeed to increased knowledge. Such advocacy appears to be uncommon and contrary to the myth that only hard facts sway the true scientist. Yet those who hold and pursue a working hypothesis are in the comparable position of sponsoring an idea on the basis of usefulness.

The so-called hard evidence is available to substantiate earth sphericity. The curvature of the earth has been measured in at least twenty different places on its surface and the results are just about the same; the curvature as seen from the artificial satellite Lunar Orbiter 1, August, 1966, was identical to the others and only a sphere has equal curvature throughout. A traveler going north from the

Courtesy: NASA

*The rising earth photographed by the Apollo 8 astronauts coming
from behind the moon, December 29, 1968.*

equator would see Polaris, the north star, rise 1° in the sky for every degree of latitude covered. If the earth were shaped like a banana, Polaris would be seen on the horizon at the equator and 90° above the horizon at the north pole, but the 1° rise for every degree traveled would be gone. Then, too, objects at the same distance above sea level fall to the earth at the same rate. One good way to interpret this fact is to imagine that the earth is a sphere.

Since the last or any so-called hard evidence can, by itself, be misinterpreted the best case for earth sphericity is had by summing the evidences. On such a basis there is only one inevitable conclusion. Making such an inductive statement does not necessitate adding all the evidences because all may never be available or be obtainable at great cost and effort. A number of pertinent evidences is enough to fortify the conclusion.

The shape of the earth is best described as an oblate spheroid. Its equatorial diameter is 27 miles more than its polar diameter, giving a small equatorial bulge.

The difference between the two diameters divided by the larger one gives a value called oblateness. For the earth the ratio is 1/297, but a more graphic picture is available by comparing the earth to a more familiar object. Were the earth to be a globe 18 inches in diameter, the difference between equatorial and polar diameters would be 1/16 of an inch; the earth would be smoother and rounder than most balls in a bowling alley, even though the earth is not the most perfect sphere in the solar system.

The size of the earth was first accurately measured by Eratosthenes (276 B.C.–196 B.C.) by noting simple events at two Egyptian cities, Cyene and

Alexandria. The same method can be given a modern cast by using Chicago and New York, about 900 miles apart. Any two fairly close places on the surface of the earth can be used. The only requirement is that the rays of the sun to both are parallel; Sao Paulo and New York thus do not fit the bill.

At noon in Chicago, when shadows are the smallest, the sun is not only the highest in the sky but is also in the center separating eastern and western halves. An imaginary line dividing the sun into two equal parts coincides with another imaginary line dividing the sky into two equal parts, the meridian. Most people know about the latter through its abbreviation A.M. and P.M. signifying ante and post meridian. Noontime in Chicago coincides with 1 P.M. in New York City, although in both places the sun's rays come in at the same angle.

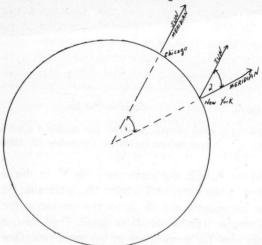

A Modern Version of Eratosthenes' Procedure

The angle at the very center of the earth, labelled 1 in the diagram, is equal to the angle between the sun's rays and the meridian at New York, labelled 2 in the diagram. The reason is the dictum that alternate interior angles formed by a line cutting two parallel ones are equal. Eratosthenes accepted without question this Euclidean postulate. Another he assumed to be true was that an interior angle at the center of a circle cuts off an arc equal in value. Angle 1 in the diagram cuts off the distance between Chicago and New York, about 900 miles. Angle 2 in the diagram can also be measured and assume it, for learning purposes, to be 8°15'43". This must be the measure of angle 1, and the following proposition can be stated: if 8°15'43" cuts off 900 miles, then an entire circle, 360°, cuts off how many miles?

$$\frac{8°15'43"}{900 \text{ miles}} = \frac{360°}{x \text{ miles}}$$

The value of x is that of the circumference of the earth, obtained quite accurately by Eratosthenes. His result was very close to our presently-accepted one of about 25,000 miles.

One of the most striking features of the procedure of Eratosthenes is the realization of his assumption that the earth is spherical. Many moderns, even the sophisticated, somehow come to the notion that the demonstration of Christopher Columbus at the end of the fifteenth century was one of the first to alert us to the belief. The truth is that a considerable number of early thinkers maintained the sphericity, or at least the curvature, of the earth.

The earliest thinkers did not realize the nature and extent of the earth's motions. A considerable number of earth movements can be documented and not all can be satisfactorily explained today. For example, Chandler's wobble, first observed in 1891 by S. C. Chandler (1846–1913) is a circular oscillation of the earth's axis of rotation taking 428 days. In 1744 mathematician and physicist Leonhard Euler (1707–1783) predicted a smaller value on the basis of his general rule for variations in a rigid, rotating object. The earth's elasticity as well as mobility of its seas may be a reason for the discrepancy between theory and observation. Chandler's wobble brings an increased latitude in one place and a simultaneous decrease $180°$ away in longitude. Thus Berlin and Honolulu have shown changes up to .3 of a second of arc.

The major motions of the earth cannot only satisfactorily be explained but also evidence for most of them is extensive. Rotation, revolution and precession are definitely in the latter category while motions in a straight line and with the galaxy are not.

Diurnal motion of all astronomical bodies, described in Chapter 2, is an evidence of earth rotation. Early peoples could view the turn of everything in the sky once every 24 hours as a real motion but in our day such circumpolar rotation is known to be a reflection of the earth's movement.

Other evidences which alone can be misinterpreted are the oblateness of the earth, the eastward deviation of falling bodies and the flow pattern of the Mississippi River. The earth's bulge can be assigned to earth rotation and in support the bulge of other planets spinning rapidly is large and those spinning slowly have a small or non-existent bulge. The eastward deviation of falling objects can be measured with sensitive instruments but like the flow of the Mississippi and some other rivers, earth rotation is cited as a cause through a circular sort of reasoning. In accordance with the manner objects attract, the Mississippi River is closer to the center of the earth near the river's head; the river should flow toward the head, northward. Since it does contrary, a cause is sought in earth rotation, ignoring factors such as gradient.

The Foucault Pendulum experiment devised about 100 years ago by Leon Foucault (1819–1869) is the most substantial evidence for earth rotation; it is not susceptible to misinterpretation. However, the nature of a pendulum and the meaning of direction must be appreciated in order to know the value.

Any pendulum, Foucault or not, has the property of maintaining a fixed plane of vibration. Once started to swing to and fro, pendulums remain in the same path unless made to change. The pendulum in a grandfather clock has this characteristic as does the simple contrivance of a piece of chalk at the end of a string. The behavior is dictated by the basic generalization called Newton's first law of motion, that things tend to stay put.

A Foucault pendulum is hung via electromagnets or a ball and socket joint to be able to swing in different paths. The pendulum is therefore called a freely-swinging one. When fastened to the ceiling and set in motion, it moves to and fro in the direction pushed, but after the elapse of some time and when still swinging, the path vibration is different. Yet no tangible force has made the pendulum shift its direction of swing. The inevitable conclusion is that the change is an illusion; the pendulum has the same direction as at the start but the earth has turned underneath.

The so-called practical person scoffs at the notion that the earth turns underneath and the pendulum is in reality in the same path. The skepticism can be short-lived by pointing to the illusion of the sun moving in our sky. Another serious objection made to the interpretation is questioning the necessity of a freely-swinging pendulum. Should not any pendulum show the same phenomenon? The Foucault pendulum is hung to be, in essence, free in space whereas an ordinary pendulum moves with the object to which it is attached.

The most convincing aspect of the Foucault demonstration comes with the performance of the experiment in various places from the north pole to the equator, or from the south pole to the equator. At the axis of the earth, a Foucault pendulum changes its direction of swing most rapidly, and the speed of movement to a different path slows as the equator is approached. At the earth's midline, a Foucault pendulum behaves like any other and remains in a fixed plane of vibration.

The reason for the gradation in speed of change is best seen with a plot of ground on earth. A section at the equator has the same north-south direction regardless of the time of day. The north-south direction shifts with time in areas away from the equator and the direction change is more rapid as a pole is approached.

The regular change in geographic direction is not only representative of what the Foucault pendulum reveals but also is evidence for the earth's sphericity. The regular curvature of a sphere is the only shape able to accommodate the feature.

If ever a Foucault pendulum were to be hung from a side wall vibrating toward the ceiling and floor, the device at the equator would then show the greatest change of path of swing. The pendulum would then indicate how east-west geographic direction shifts with time. Such an imaginary pendulum would have slower changes in path of travel as a geographic pole is approached.

Hanging such a pendulum seems as unlikely as anyone ever directly sensing the rotation of the earth. The speed of $360°$ every 24 hours, or $15°$ an hour, is had by all occupants of the earth at all times. The varying linear rotation speed also cannot

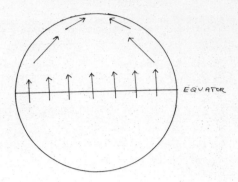

How North-South Direction Changes

be directly observed. Equatorial points turn at the rate of about 1000 miles an hour but the smaller circles towards the poles have slower linear speeds. At New York City, the rotation rate is 12¾ miles a minute.

The interior of the earth whether deep in a mine or a portion of the core has identical angular speed, 15° an hour, but linear speed varies with the distance from the axis. Places closer to the imaginary line about which the spin occurs have a slower linear speed.

The contention that only the crust of the earth is rotating and the remainder is stationary cannot be tolerated. Such a planet, comparable to a stationary grapefruit and a rotating peel, would generate a great deal of heat in friction not now detected. Moreover, if ever a digging tool entered the region of supposed immobility, the device would break at the junction of rotating and non-rotating portions of the earth.

Shifting between various portions of the earth does affect the rotation speed. It is one of many factors causing small irregularities in the rate of turn about the axis. Other causes include temperature changes and movements in the solid crust but the most significant factor is the friction between the oceans and the solid crust. The irregularity is not very much causing a change in the length of day of about .0016 seconds a century.

The flow of water is a close rival of the Foucault pendulum in demonstrating earth rotation. Under perfect laboratory conditions, water draining in the northern hemisphere swirls counterclockwise while in the southern hemisphere, the water let out of a sink swirls clockwise. If the earth were not rotating, the deflection would not be observed. The water, however, in the northern hemisphere is moving from a place of lower linear rotation speed to one with larger linear rotation speed and appears to be deflected to the right. The effect, the Coriolis force, is named after the French engineer and mathematician Gaspard G. De Coriolis (1792–1843) who first described it.

The earth's revolution is also indirectly experienced. The average rate is about 18½ miles a second, or 66,000 miles an hour. Those who lead a fast life as well as the sedentary are trapped in this rate but yet have no direct sense perception of the speedy motion.

The evidences for earth revolution are substantial; all intelligent persons today believe that the earth turns about the sun. Indeed the success of the idea in explaining and predicting can be cited as evidence. The observational type of support is also available.

Throughout the year a different group of stars rises in the east as the sun sets in the west. This phenomenon must be adjudged as a reflection of the earth's travel about the sun; else the outmoded idea that the situation is reversed must be reconsidered. It is not reconsidered because stellar parallax and the aberration of light, among other facts, zero in veritably to prove the earth's travel about the sun.

Stellar parallax was vainly sought for several generations after Nicholas Copernicus (1473–1543) reintroduced the idea that the earth and other planets went about the sun. Critics of the conception could rightfully ask why the near stars were not seen in varying perspectives among the more distant stars. Isaac Newton (1642–1727) could argue that all the stars were much too distant to show the effect even though his estimates were far from the present day measurements of star distance.

What was sought can be demonstrated by holding a finger in front of the nose and observing the place of the finger in the background with one eye at a time. The finger represents the near star and each eye would be the earth in two extremes of its orbit, six months apart.

Stellar parallax was finally detected in 1838 by Fredrick Bessel (1784–1846) because much finer telescopes became available. Two other astronomers, Thomas Henderson (1798–1844) at the Cape of Good Hope in Africa and Fredrick Georg Wilhelm Struve (1793–1864) in Russia also measured stellar parallax in the same year. Since then the parallax of about 700 stars has been recorded.

The aberration of light was found much earlier, by James Bradley (1693–1762) at Oxford University. When sailing, he found the weather vane on the mast of the boat shifted its position each time the boat changed its course, although the direction of the wind was constant. The weather vane took a heading due both to the wind and the motion of the boat. In a similar manner the position of a star is determined by the direction of the light from it and the motion of the earth.

Another analogy to the aberration of light is had in the experience of holding an umbrella. When standing in the rain, a person holds the protective cover upright but when walking or running, the umbrella is tilted. Similarly telescopes must be tilted, and at varying angles, as the earth moves in one direction and then the other about the sun.

If the earth were not moving in space, telescopes would have an invariable direction when pointed at a star. However, telescopes must be pointed at slightly different angles throughout the year. Each star appears to have a small ellipse of position in any one year that is a reflection of the earth's motion about the sun.

Another major motion of the earth can also be detected as a reflection in the sky. In contrast to rotation and revolution, precession, or precession of the equinoxes, is very slow; one complete cycle takes 26,000 years. During that time the change in axis direction shows as a variation in our pole stars.

The earth's axis is said to point to Polaris but actually it is about a degree away from the star. Hundreds of years ago the axis pointed to Er Rai; the star Thuban was also once the north star. In about 13,000 years, the star Vega will be the north pole star. Twenty-six thousand years from now, Polaris will again be near where the axis is pointing.

Precession was first measured by the father of Greek astronomy, Hipparchus (180 B.C.–110 B.C.). He was able to see the shift at the horizon, noting the variation in the stellar background from year to year when the sun rose due east. His value was 40 seconds of arc per year, compared to today's 50.3 seconds of arc per year. At the time of Hipparchus, the constellation of Aries rose with the sun on March 21. Earlier, the constellation Taurus was in this position. Today, Pisces is the group involved. That Hipparchus was able to make a numerical description of this shift attests to the reason for his prominence in antiquity.

The cause of precession is the gravitational pull of the sun and to some extent the moon on the bulge of the earth. The attraction results in the wobbling, slow change in axis direction, opposite to the eastward motion of rotation and revolution.

The sun is the original motive power for the earth's rotation and revolution, and with the moon is the cause of earth precession. Our sun is also involved in several other earth motions.

The distance between the sun and the constellation Hercules is shortening at the rate of 12 miles a second. The earth and all other parasites of the sun must also have the motion, if indeed the sun moves in that fashion. The earth's elliptical orbit based on a stationary sun must become a spiral in space. If the sun is moving in a straight line at the speed of 12 miles a second then the earth's spiral in space is neither very compact nor loose. The earth's forward motion of 12 miles a second is paired with an average revolution motion of 18½ miles a second, making a spiral with an approximate 2 to 3 ratio in dimensions.

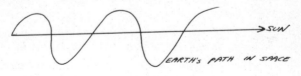

A Real Path of Earth

The sun's motion towards Hercules may be a part of some other since all visible stars, including the sun are turning regularly in the tremendously large pattern of stars called the galaxy. It is unlikely that the sun or the stars in Hercules would deviate. Therefore the motion could be a part of the sun's revolution within the galaxy. Our star has the enormous speed of about **160** miles a second because the entire galaxy is spinning. The earth and all on the planet partake of this motion but it has minimal effect on the living and lifeless. Every particle of the earth is attuned to the speed continually and only a change will make an effect.

Of all the earth motions, rotation and revolution have a prime influence on the living and lifeless. The sequence of day and night is important to all while the earth's travel about the sun is a cause of the seasonal cycle. The other major cause of the seasons is the inclination of the earth's axis.

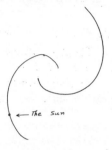

← The Sun

Our Spinning Galaxy

The inclination of the earth's axis is glibly given as 23½°. The information is better conveyed by a diagram because an angle involves two lines and only one is directly given in the statement. The sketch reveals two angles that are 23½°. One is between the axis and another imaginary line perpendicular to the orbit of the earth and the second is between the equator and the orbit of the earth.

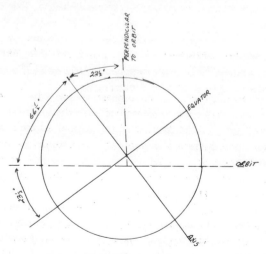

The Inclination of the Earth

The inclination of the earth could just as glibly be said to be 66½°. The angle between the axis and the line representing the orbit of the earth is such a value.

No alternative is available, however, for the fact of inclination. If the earth were not tipped, the seasonal cycle on earth, the periodicity in place of sun rise and sunset, and precession would be altered. An earth with an axis coinciding with the perpendicular to the orbit would not have precession. The inclination is the chief reason for the seasons as well as the accompanying position of the sun in our sky.

The supposition that the orbit is tipped rather than the axis is not a tenable alternative; precession could not occur. A sequence of seasons would be maintained.

Winter in the northern hemisphere occurs when the earth is closest to the sun, about 91½ million miles away. The rays of the sun do not strike the northern hemisphere directly; instead they are spread over an area with the sun being low in the sky. During the northern hemisphere summer, the sun is about 94½ million miles distant but the sun's rays hit the earth more directly.

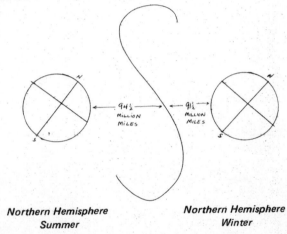

Northern Hemisphere
Summer

Northern Hemisphere
Winter

The southern hemisphere has its summer season when the earth and sun are closest. Yet the three million mile difference with the position for the northern half of the earth does not give the southern half warmer summers. The southern hemisphere has more water surface and water acts as a moderating influence in climate. The southern hemisphere has its winter when the earth-sun distance is farthest but the winters are not any more severe than those in the north; the larger percentage of water is again the moderating factor.

In about 13,000 years, after half a cycle of precession, the seasons on the earth will shift. The northern hemisphere will have its summer season when the earth and sun are closest and its winter season when the earth and sun are farthest; summers should be slightly warmer and winters slightly colder.

SELECTED REFERENCES

Guy Murchie, *The Music of the Spheres*, 2 vols., New York: Dover, 1967.
George Gamow, *Planet Called Earth*, New York: Viking, 1970.

CHAPTER 4

THE MOON

As seen from the earth, the disk of the full moon appears to be equal to the disk of the sun. Both have an angular diameter of about 31'. Oddly, both sun and moon have an average distance to the earth of about 110 times their respective linear diameters. For the sun, 110 times its diameter, 864,000 miles, yields 95,040,000; for the moon, 110 times its diameter, 2,160 miles, gives 237,600 miles.

Long before man first landed on the moon, July 20, 1969, many students of the moon, selenographers, had prepared huge maps and globes of our natural satellite. In 1924 the British amateur H. P. Wilkins (1896–   ) built a model of the moon 60 inches in diameter; in 1932 he had one 200 inches in diameter; in 1945 he had a globe of the moon with a diameter of 25 feet. The third edition of the moon map sold through the Association of Lunar and Planetary Observers as well as the amateur's journal, *Sky and Telescope*, was in 25 sections each 20 by 21 inches. A 40-foot mural of the moon was in the lobby of the Boston Planetarium.

The National Aeronautics and Space administration constructed a map of 144 pieces, each 22 by 29 inches. During the preparation for and the actual flights to the moon large numbers of photographs were taken. Our natural satellite is now much better charted.

The moon's surface features, first delineated more carefully by Galileo (1564–1642) when he published his use of the telescope, are now known through modern instruments. Craters, maria and other geological features can be minutely described.

The craters on the moon are immense compared to those on earth. The meteor crater in Arizona, 19 miles west of Winslow and 40 miles east of Flagstaff, is about a mile in diameter; a crater now filled with water in northern Quebec is about twice the size. The smallest on earth may be the Carolina Bays, along the Carolina and Virginia coasts, seen as depressions from aloft but not when walking over them. In contrast, one of the largest on the moon, Clavius, is about 150 miles in diameter.

Many of the thousands of craters on the moon have been named; several hundred of them have been precisely measured. Since the names were largely given during earlier centuries, few United States citizens have craters named in their honor. Two women, however, are so memorialized.

At the end of the eighteenth century, J. H. Schröter (1745–1816) found that for each of the many craters he measured, the part of the material above the surface is approximately equal to the volume of the interior depression below the surface. He and others noted how the craters were more like saucers than cups. Moon craters are extremely shallow in proportion to their width.

Seen with even weak binoculars or a cheap telescope, craters can be a truly impressive sight. For example, the triple crater, Theophilus, Cyrillus, and Catharina, best viewed when the moon is 5.5 to 7.5 or 18 to 20 days old is such a spectacle. Catharina, the largest of the three is about 70 miles in diameter. Theophilus is the deepest, about 18,000 feet, with a central mountain peak rising about 6,000 feet from the floor. From Tycho, the so-called "metropolitan crater of the moon", radiate out a conspicuous system of light streaks or light rays.

The American explorers of the moon examined craters directly as well as through photographs at close range. Lunar Orbiter photographs showed small, perfectly circular craters and deep "fresh" ones. A very old rock familiarly called the "genesis rock" was picked up near Spur Crater.

The exploration did not change the theory that craters are due to bombardment, volcanic activity and/or natural cooling. The meteor bombardment idea can account for some features, notably random distribution. Areas of cinder cones, dead volcanic craters, supports the volcano origin thesis. Watching melted chocolate cool is an appealing support for the hypothesis that craters appear in a similar fashion. Of course, the moon is made of green cheese, not chocolate!

Whoever fashioned the fable about the cheesy nature of our natural satellite did not look very closely. With unaided eye, many dark areas called maria can be seen. The ancients named many of those, although their Latin word for sea, maria, is as erroneous as is the green cheese. Mare Crisium, sea of crises or conflicts, is one of the largest, being 70,000 square miles in area. Among other appealing names are Mare Nectaris, sea of nectar, and Mare Serenitatis, sea of serenity.

The maria regions cover about one-third of the near side of the moon and are a smaller fraction of the far side. They are the most recent widespread rock formation on the lunar surface. The first samples of rock taken were from these regions. Early in February, 1971, samples were taken from Mare Imbrium, the sea of showers. The total collection then included 33 rocks weighing more than 50 grams each and about 30 smaller rocks weighing from 10 to 50 grams each. Analysis of these basaltic rocks showed some chemical composition differences from those first acquired.

The manned lunar investigations have revealed a moon with a warm interior and chemically complex. Some moon rocks have almost no strontium and europium; others have large amounts of these chemical elements. Rocks from the eastern maria have much more titanium than those from the western maria.

The Apollo missions 11 through 17 (except for 13 which was aborted) resulted in about 800 pounds of moon material brought to the earth. Unlike the government-sponsored Wilkes expedition (1838–1842) of wooden sailing ships for cartography exploration, where results were not published, the National Aeronautics and Space Administration widely disseminated their findings, and some reinterpretations occurred.

Geological features other than craters and maria abound on the moon. Some of the mountain ranges are named after those on the earth so that reference to the Carpathians, Alps, or Apennines, among others must in the future be more specific. During their three-day stay on the lunar surface two astronauts during late 1971 discovered layering on the inside walls of the mile-wide gorge known as Hadley Rille. Less-visible geological features announced in 1968 were mascons, mass excesses in the circular maria regions. The first interpretation of these anomalies hold remnants of colliding iron asteroids, buried under the lunar surface, to be responsible.

The geological features of the moon like those of the earth are subject to change although the average person believes that the moon is a dead world. An array of impressive evidence is available to show that some geological surface features have altered with time.

Two craters seem to have disappeared. At the beginning of the nineteenth century, a competent moon observer cited one near the border of the Mare Crisium and it is not seen today. The crater Linné, about six miles in diameter, disappeared sometime between 1843 and 1866; afterwards a white spot surrounding the crater diminished in size and brightness before growing again.

At the beginning of the nineteenth century Johann von Mädler (1794–1874) of Berlin described a remarkably perfect square on the region of the moon between the ring plane named Fontenelle and the walled enclosure called Birmingham. He said he saw a regular cross on the floor. In 1876 the director of the Natal Observatory in South Africa described essentially the same structure. In the middle of the twentieth century, one amateur failed to see it while Patrick Moore (1923–   ) in Britain saw three of the four walls forming the square.

Photographs of the moon made on October 26, 1956 with the 60-inch reflector at Mt. Wilson Observatory revealed an "obscuration" over the crater Alphonsus. The soviet astronomer, Nikolai A. Kozyrev (1908–1958), on November 3, 1958 detected a discharge of gas over the crater; two weeks later a cloud was seen over Alphonsus by astronomers in England and the United States.

During the early 1960's, astronomers noted transient luminous red spots on the moon; at Lowell Observatory, Flagstaff, Arizona, three red spots lasting less than 20 minutes were detected on the rim of the crater Aristarchus. However, all such examples of moon luminescence have been traced to solar activity. Powdered samples of material similar to moon rocks emit red light when struck by energetic protons of the kind yielded by the sun.

The changes on the moon may prompt an interpretation at variance with the basic precepts of science. Any introduction of the supernatural, for example, cannot be tolerated. The scientific enterprise deals only with natural phenomena and natural causes.

The moon hoax perpetrated by a New York newspaper at the beginning of the nineteenth century was not, however, in the category of bringing in the supernatural. The *New York Sun* in 1835 simply deceived their readers with false reports from a South African observatory about life on the moon. The astronomer in charge denied the contentions but during the months his message was in transit the newspaper tripled their circulation.

Substantiated facts are the basis for scientific judgments. This is why the conclusion that the moon lacks an atmosphere was well corroborated long before man set foot on our natural satellite.

The evidence for the lack of lunar atmosphere is substantial. There are no clouds nor storms on the moon; there is no haze because all shadows are black. Parts near the edges of the moon are seen sharply; no twilight occurs at the cusps of the crescent moon. When the moon begins to block a star's light in the phenomenon called occultation, the starlight never pierces a lunar atmosphere.

The moon does not have the gravitational attraction to be able to hold a blanket of gas. If an atmosphere ever did exist about the moon, the gas would need only a speed of 2.38 kilometers per second to escape. The escape velocity of the earth is considerably more at 11.18 kilometers per second; that for the sun is 618 kilometers per second.

The spatial relationship between the earth, moon and sun accounts for several effects on the earth as well as the manner in which the moon is seen on earth. Eclipses, tides and phases of the moon are controlled by where the three objects are with respect to each other.

If a straight line can be imagined drawn through sun, earth and moon, in that order, then a lunar eclipse is probable. Both earth and moon cast conical shadows while the sun, a self-luminous body, has no shadow. The earth's shadow is large enough to cover the moon, making a lunar eclipse.

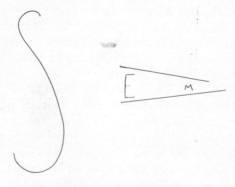

Condition for a Lunar Eclipse

If a straight line can be imagined drawn through sun, moon and earth, in that order, then a solar eclipse may be seen in certain sections of the earth. The solar eclipse may fail to occur for two reasons. First, the moon's conical shadow may not be long enough to reach the earth. More often, the moon and earth are in slightly different planes with the shadow of the moon stretching above or below the earth.

A lunar eclipse can be enjoyed without thought of eye damage. A solar eclipse, however, is deceptive in that the sun appears to be shaded, but in reality the rays are as dangerous as ever. Observers have been blinded by foolishly staring at the phenomenon. Early records describe solar eclipses without mentioning the danger

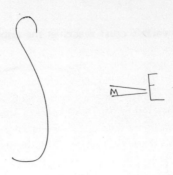

Condition for a Solar Eclipse

involved. In the book of Amos in the Bible is: "And it shall come to pass in that day, saith the Lord God, that I will cause the sun to go down at noon, and I will darken the earth on a clear day." In Homer's *Odyssey*, Ulysses after his wanderings returned home and killed the suitors of his wife: "The sun has perished out of heaven and an evil mist has spread over all." There was a total solar eclipse around Ithaca on April 16, 1178 B.C.

Eclipses can be predicted far ahead and calculated for exact time of occurrence in the past. During the nineteenth century, a European astronomer tabulated eclipses in this manner between 1207 B.C. and 2162 A.D. It is not a modern accomplishment because legend would have a king of China 39 centuries ago punishing the astronomers who indulging in drink failed to foretell an eclipse.

Total eclipses of the sun have instigated much human drama. The one occurring May 28, 585 B.C. made the Medes and Lydians stop their six-year war in Asia Minor. The event on June 20, 840 brought the death, perhaps of fright, of Charlemagne's son, Louis of Bavaria. The solar eclipse sighted in Uruguay, May 29, 1919, brought substantiation for the General Theory of Relativity. The June 30, 1973 event lasted for 7 minutes and 3.9 seconds, longer than all but two other solar eclipses during the last 1,433 years.

Whereas both lunar and solar eclipses are determined solely by the configuration of earth, sun and moon, tides have additional causes besides the arrangement in space of the three objects. Terrestrial influences on water tides include the size, depth, density and natural rhythm of the body of water; other controlling factors are the friction between ocean and land as well as atmospheric pressure.

Tides are associated with water because of the visible evidence and economic importance. However, atmospheric tides exist but have not been studied and land tides may be responsible, for example, for the omnipresent cracks formed in buildings. At least in midwestern United States the tidal rise and fall of the solid surface may be as much as one foot.

Daily and monthly tides occur and in both instances the chief astronomical cause is the moon. A daily high tide occurs every 12 hours at any given locality, when the place is directly under the moon. Another daily high tide occurs about

180° away because the earth's crust reacts in the opposite direction in its thrust towards stability.

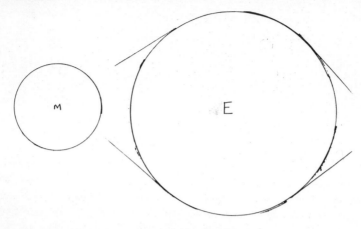

The Daily High Tide

Monthly tides are due primarily to the difference between the moon's pull for the whole earth and the moon's pull for the surface of the earth. The first value is given in accordance with the law of gravitation by the product of the masses of the earth and moon divided by the square of the distance between their centers. The same formula holds for the pull on the surface of the earth but in this case the earth's mass is considerably reduced and the distance is smaller by the earth's radius, 4,000 miles. The sun has less effect on the production of tides although its attraction to the earth as a whole is much greater than is the moon's. The difference in the sun's pull on the whole earth and the sun's pull on the earth's surface is governed by the same decrease in the earth's mass and the identical change in distance between centers. However, 4,000 miles is hardly a dent in 93,000,000 miles while the earth's radius make a sizeable change in 239,000 miles. The sun's much greater distance prevents the same amount of differential pulling done by the moon.

The sun does make a contribution to earthly tides. When a straight line can be imagined between earth, moon and sun, the surface of the earth is raised most by the forces of attraction. The situation is similar whether the line of objects is earth, moon and sun or moon, earth and sun. In both cases, the rise of earth's crust brings an almost identical rise about 180° away. It is much like the lady and her child waiting in the railroad station. She holds the child by one hand while his other is free to move. The moment she runs to meet the train and tugs at his hand, the relatively free hand jets out to maintain balance. In any event when the three objects are in line, the monthly tides, called spring tides, are highest.

When two straight lines can be imagined connecting earth, moon and sun, particularly two straight lines perpendicular to each other, tides are smaller. The smallest monthly tides, neap tides, occur when the sun, moon and earth can form a right angle.

Conditions for Spring Tides

Conditions for Neap Tides

Because many non-astronomical factors influence tides, their prediction is difficult. Observation rather than calculation reveals the tidal range at the Atlantic end
of the Panama Canal is 1 to 2 feet while at the Pacific end, only 40 miles away, the
range is from 12 to 16 feet. The highest tides, about a fifty-foot rise, are in the Bay
of Fundy, between New Brunswick and Nova Scotia, Canada. At least 5 other
places have a tidal range of more than 30 feet: Puerto Gallegos in Argentina; Cook
Inlet, Alaska; Frobisher Bay in Davis Strait, Northwest Territories, Canada;
Koksoak River into Ungava Bay, Northern Quebec, Canada; Gulf of St. Malo,
France.

Tides on the moon have not been fully investigated. The ebb and flow of the
lunar crust should follow the same kind of pattern as the rise and fall of the earth's
crust. The moon's daily and monthly tides would appear to be smaller than the
earth's but non-astronomical factors on the moon need to be more carefully
charted. One effect of tidal friction, operating for billions of years, may have been
to heat the moon's interior. The side of the moon facing the earth has a permanent
bulge of about 165 feet that increases about 10% in size when earth and moon are
closest and decreases when the two are most distant; this tidal oscillation causes
heat and other effects.

The moon's revolution about the earth can be considered in several ways. First,
the moon can be considered moving about the sun and disturbed by the presence of
the earth. Second, the motion can be viewed as the monthly revolution around the
center of mass of the earth-moon system, and the sun is a disturbing factor. The last
is the most popular conception: The moon has an elliptical orbit about the earth.
The first view is supported by charting the path with respect to the sun and the
moon's path is always concave to the sun. The second has a basis in the fact that
the earth and moon are an unusual system; no other planet has a close moon as
large. The last is most practical for daily affairs.

The time for the moon to travel about the earth can be measured several different ways. Two successive views of the moon when closest to the earth, the perigee
position, gives the anomalistic month of about 27-1/2 days. When the point of
intersection between the earth and moon orbits are used, the nodes, the nodal or
diaconic month is about 27-1/5 days. The duration between successive times the
moon, earth and sun can be considered in one straight line is called the synodic
month. It is about 29-1/2 days. The sidereal month is about 27-1/3 days and is the
time for two successive sightings of a distant star in line with the earth and moon.

The sidereal and synodic months vary in length because the moon and earth do
not travel much in space with respect to a distant star. In the synodic month, the
earth, and moon, move with respect to the sun, and an extra two and a fraction
days are required for alignment.

The synodic month must be used in adjudging phases; the moon is visible because it reflects sunlight. During the 29½ day period the moon covers approximately 360° so that 360° divided by 29½ or about 12° is covered every 24 hours.

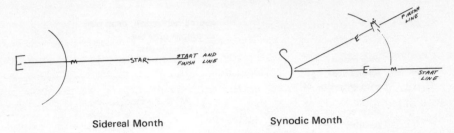

| Sidereal Month | Synodic Month |

Sidereal and Synodic Months

When sighting the moon in any sky at a given time during two successive days, the 12° drift is easy to discern. The moon has the largest eastward drift of all. As indicated in Chapter 2, even a youngster can make out the large shift of the moon from one day to the next.

The 12° eastward drift can be translated into its meaning for time by considering the basis of ordinary time, 360° every 24 hours. Fifteen degrees is equivalent to one hour and 12° is about 51 minutes. Therefore the moon rises about 51 minutes later each day.

The full moon rises when the sun sets and the full moon sets about 12 hours later, when the sun rises. If the full moon rose when the sun set, say six P.M., then the next day, the slightly-less-than-full moon rises at 6:51 P.M. The second day after full moon, the rise of the moon will be about 7:40 P.M. About seven days after full moon, the moon is seen as half a circle, rising about midnight and setting twelve hours later at noon. The moon continues to show a smaller and smaller circle as a waning moon, rising later in the night, for about seven days after waning half moon. Then, approximately 14 days after full moon rose with the sunset, the new or no moon rises with sun and sets with the sun. The waning moon can be associated with darkness although the crescent moon rising after midnight is also seen during the day.

The waxing moon can be associated with daylight hours because from immediately after new moon to full moon, its time of rise is after the sun does. At about noontime, a waxing half moon rises and sets about 12 hours later, at midnight.

Since the moon is a day phenomenon approximately half the time, romance and poetry on earth should flourish day and night — unless the full moon is the necessity for the activities.

The full moon seems to yield a good supply of illumination. The albedo, the ratio between the amount of light reflected and the amount that falls on the object, is .073. Of course, the albedo varies on different sections of the moon and .073 is the overall average. The meaning of the number can be visualized by comparing the full moon and sun. If the entire visible hemisphere of the earth were packed with full moons, the earth would receive only about 1/5 of the light of the sun. A half-moon should appear to be 1/2 as bright as a full moon, but actually the half moon is 1/9 as bright. The reason is that the light of the sun strikes obliquely at the line separating the dark and reflecting portions of the moon, the terminator.

M² waning half moon, neap tide
no eclipse possible
rises at midnight

Sun
new moon,
Spring Tide,
rises with sun, M³

solar eclipse
possible

E

M¹ full moon, spring
tide, lunar eclipse
rises when sun sets

M⁴ waxing half moon, neap tide
no eclipse possible,
rises at noon

Tides Eclipses and Phases

The full moon appears to show half of the surface of our natural satellite. A careful measurement at any one time would reveal about 41% of the total surface, and year after year the identical surface is seen by earth observers. The equality of moon rotation and revolution speeds, about 1/2 mile per second, is the reason for the effect.

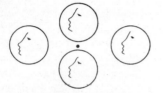

*Rotation Period Does Not
Equal Revolution Period*

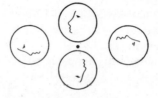

*Rotation Period Equals
Revolution Period*

A variety of Moon tipping as well as a strategic view from the earth increases the percentage of moon that earth observers can see to about 60%. Librations are the name given to all of the effects. Libration in latitude yields about 6½° beyond the poles of the moon, because the lunar poles are tipped alternately toward and away from us at intervals of two weeks; the moon has an inclination of about 6½° between its equator and the plane of its orbit. Libration in longitude is due to the failure of the moon's rotation and revolution to keep exactly in step through the month although they come out together at the end. The rotation is uniform but the revolution is not because the moon's orbit is an ellipse. The moon consequently has an east-west rocking allowing earth observers to see as much as 7¾° more at each edge. The moon is not a perfect sphere and physical libration results. Diurnal libration is not due to the moon but to the earth's rotation and allows earth observers to look over edges of the moon about 1° more when the earth is rising and setting.

In October, 1959, the first sight of the back of the moon, the remaining 40 percent, was televised to the earth by a Soviet lunar spacecraft. More photos of this region confirms the prediction made by the principle of the uniformity of nature that the rear side of the moon was essentially like the side seen by observers on the earth.

Long ago earth observers in an agricultural society used the first full moon after September 21 as a guide to harvest and to this day that full moon is called the harvest moon. The next full moon is called the hunter's moon.

Ancient observers also noted that a full moon seen on the horizon appears larger than the same full moon seen on the midpoint of the sky. This moon illusion, disturbing to some today, was first adequately explained by Claudius Ptolemy (100–170 A.D.). He claimed that any object seen through filled space, such as the moon on the horizon, is perceived as being more distant and larger for mostly psychological reasons.

A full moon does not occur in every February. The month was without a full moon in 1885, 1915, 1934, and 1961. In 1866, February had no full moon while January and March each had two full moons; this will not occur again for 2,500,000 years.

When the moon is full its angular distance from the sun, elongation, is $180°$; this is the opposition position. When the moon's elongation is $0°$, this conjunction position means a new moon. A word suitable for crossword puzzles, syzygy, is a suitable description for the opposition or conjunction positions. A half moon has a $90°$ elongation and the position is called quadrature. These terms for angular positions are also suitable for artificial satellites.

The possibility of another natural moon has intrigued some theoreticians since the time of French mathematician and astronomer Joseph Louis Lagrange (1736–1813). In 1772 he calculated that there are five points of gravitational equilibrium around a pair of massive bodies; three represent an unstable equilibrium and two are gravitational points where a small body tends to stay. In 1904 a small mass was found in a point of the sun-Jupiter system and it proved to be an asteroid, a small planet between Mars and Jupiter. In the 1950's a Polish astronomer looked for similar objects in the earth-moon system. He found two faint clouds, probably collections of meteoritic material, circling the earth at one of the points.

Observers on the moon will have much better viewing conditions to detect any possible natural satellite. However, a human visitor to the moon will need to bring much scientific equipment as well as a supply of food, water, air and heat. Unless the visitor can move about all portions of the moon, heat is necessary to survive the long moon period when the sun is not shining. The moon has no atmosphere to act as a heat reservoir; the moon is baked during its day and almost frozen throughout its night. Phenomena associated with the earth's atmosphere such as rain, snow, winds, rainbows and brilliant sky colors are not on the moon. On a harsh, grim landscape the moon tourist sees a black sky with a bright yellow sun and brilliant pinpoints of light, the stars. The motions in the moon's sky due to the moon's rotation (and revolution) are about 27 times as slow as is the earth's diurnal motion.

In 1970 two senior scientists at Soviet Academy of Science proposed what they called a crazy theory: the moon is an abandoned spaceship parked in earth orbit over two billion years ago. They claimed their idea accounted for the moon's density being smaller than the earth's because the moon had a hollow structure. Meteors strike without making a deep crater because the armor plated hull covered with loose surface soil is a protector. Craters floors are convex rather than concave because the armor hull is round. The crazy theory did not gain support, only jocular small talk about finding the door and archeological expeditions to the moon.

SELECTED REFERENCES

H. S. F. Cooper, *Moon Rocks*, New York: Dial, 1971.
G. Gamow and H. G. Stubbs, *Moon*, New York: Abelard-Schuman, 1971.

CHAPTER 5

THE SUN

The brightest object in the sky, the sun has been worshipped and respected. Modern man can best give homage to this most important star by studying and enjoying it but never gazing intently at it for a long period of time. Staring at the sun, whether or not obscured by an eclipse, is a simple way to cause blindness.

Being enchanted with the sun can best be safely accomplished by trying to see its green flash at sunrise or sunset. Beginning in the nineteenth century competent observers began to describe the streak of green seen at the edge of the sun at dawn or dusk. Not everybody who has tried has seen the phenomenon and those who have observed the glimmer of green color do not agree on the reason for it. The most popular conception is that the green flash appears after dispersion, scattering and absorption of sunlight by the earth's atmosphere; the lower, red rays have sunk below the observer's horizon, orange and yellow are largely absorbed by the atmosphere and blue and violet rays are scattered by particles in the air.

The moon, Venus and Jupiter have also exhibited the green flash but it is most common for the sun. During the June 1, 1973 solar eclipse, hundreds saw it for the sun, while on the cruise ship *Canberra* making a special tour to view the eclipse.

Being amazed by the sun can also be accomplished through reflection about its properties. For one, imagine a tremendously hot sphere, with a diameter more than 100 times that of the earth; the sun is 864,000 miles in diameter. The amount of material in the sun is 333,420 times that in the earth; the space occupied by the sun is approximately 1,300,000 times that taken up by the earth. At the very center of the sun, the material is thought to be so compressed to make the density about 80 grams per cubic centimeter; the densest element on earth, osmium, has a density of about 22 grams per cubic centimeter.

The sun is more varied than earth in density. Overall, the earth's density is 5.5 grams per cubic centimeter while the sun's is 1.4 grams per cubic centimeter. Consequently, the sun has regions like its compact center as well as places where almost nothing is present.

The photosphere is as far into the sun where instruments have been able to probe, yet the photosphere is much closer to the edge than to the center. The average temperature of the photosphere is about 5700° centigrade and it is speckled with black spots, sunspots, many hundred degrees lower in temperature than the photosphere.

Galileo (1564–1642) was the first to call attention to the sunspots yet their systematic study is only about a century old. The observatory at Greenwich, England began recording sizes and numbers of sunspots in 1874. In 1947 a spot about 40 times the size of the earth was observed, one of about 27 more than three million square miles in area.

Sunspot activity unequalled in two centuries was recorded during September and October, 1957. The so-called sunspot number — a measure of their number and size — was 244 for September and 263 for October. The previous high number was 239 for May, 1778.

The rise and fall of sunspot activity was uncovered in 1825 by H. Schwabe (1789–1875); at the time he was looking for another planet near Mercury. He established that the number and size of sunspots increases every 11 years. Before and largely after this discovery, men, women and children have tried to correlate sunspot activity with frequency of war, stock market prices, baldness in males, and a great variety of other personal and cultural patterns. No one can be prevented or ridiculed out of making such relationships; stranger ones have been verified. In this case, however, the attempt to relate sunspots and human activities is an aspect of the fraud called astrology.

Any two items may be related but only those with a high positive correlation have any meaning. The size of your breakfast and the atmospheric temperature in Nairobi, Kenya may be connected; only statistical analyses will reveal the bond. If the correlation is positive, the scientist then seeks a better pattern of interdependency.

Sunspots and radio transmission as well as sunspots and precipitation have been successfully correlated. The next step of a deeper understanding is not available although guesses have been circulated. Some see a clue in the magnetic nature of the spots and others point to the increased amount of solar material, notably protons, reaching the earth.

The sun's rays are tangible but a more substantial material emanation from the sun can be seen in the prominences. This flame-like matter is ejected from the chromosphere, closer to us than is the photosphere, seen as a bright red rim during a total solar eclipse. The most frequently seen prominences, called active, originate in the photosphere and follow a curved path back into the sun. Eruptive prominences occur in sunspot regions and may rise several hundred thousand miles before being pulled back into the sun; some separate and rise to great altitudes while fading. Other types of prominences are called tornado, quiescent, and coronal. The latter originates in the atmosphere of the sun, the corona. Early in the twentieth century the corona was thought to be the region immediately about the sun. The corona was studied during a total solar eclipse until Bernard Lyot (1897–1952) in 1930 devised a coronagraph, a disk arrangement in a specially-designed telescope capable of artificially blocking the sun to reveal the corona. Temperature measurements revealed the amazingly high temperature of one million degrees. The photosphere closer to the hot interior of the sun was only 5700°C while the corona, farther out, had a multiple of the temperature. The explanation then devised and still accepted is that the heat coming from the interior produces shock waves at the outer extremities of the sun. Water waves hitting a breakwater have the same effect but not to the same degree.

The shock wave theory is accepted today even though the corona is now considered to be veritably the entire system. The region near the chromosphere is denser and acts as a barrier to the heat flow. Perhaps such a gate is a boon to life on earth; a greater supply of heat energy reaching the planet would not be conducive to plants nor animals. The earth receives two calories per square centimeter per minute from the sun. Its quota of solar energy is about one part in two billion; in this respect, the earth is like a half dollar coin in a circular field one mile in diameter.

Each second the sun yields a huge amount of energy, the equivalent of four million tons of matter. At the very nominal cost of one cent per kilowatt-hour, several billion billion dollars would be necessary to keep the sun going for one second.

The manner in which the sun manufactured its huge supply of energy has intrigued many thinkers. During the nineteenth century scientists saw the process as one of combustion or compression on a grand scale. A more feasible idea was born with the rise of nuclear physics; at least the conception is in agreement with the measured output of the sun.

Many of the known atomic nuclei have been detected on the sun but only a few are directly involved in the nuclear synthesis process yielding energy. (Practically all of the material, about 99.996 percent, is made up of only 15 elements. The least abundant one, nickel, is more common than all the others combined.) One school of thought holds that nuclei of hydrogen atoms fuse to form helium nuclei. A simplified schema of this process does indeed reveal a tremendous manufacture of energy. Another school of thought believes that the hydrogen nuclei do indeed fuse but that carbon, nitrogen and oxygen nuclei aid the process. A third view is eclectic and assigns percentages to the two methods; some say that only 10% of the solar energy is produced by the so-called carbon-nitrogen cycle and the remainder is due to the union of hydrogen nuclei.

The settling of the problem of how the sun manufactures its energy is significant on two bases. First, other stars can be better understood. The sun is an average star and any description for it can be used with some confidence for other stars. Second, and more meaningful, is the possibility of understanding the process so well to be able to make artificial stars. Already man has started in this direction with the hydrogen bomb; the weapon is a nuclear synthesis. A controlled nuclear fusion able to be tapped continuously for energy will put man's age-old problems into history. A millenium for all peoples can truly occur.

SELECTED REFERENCES

Karl Kiepenheuer, *The Sun*, Ann Arbor: University of Michigan Press, 1959.
Donald H. Menzel, *Our Sun*, Cambridge, Mass.: Harvard University Press, 1959, Rev. Ed.

Four Hydrogen (H) Forming Helium (He)

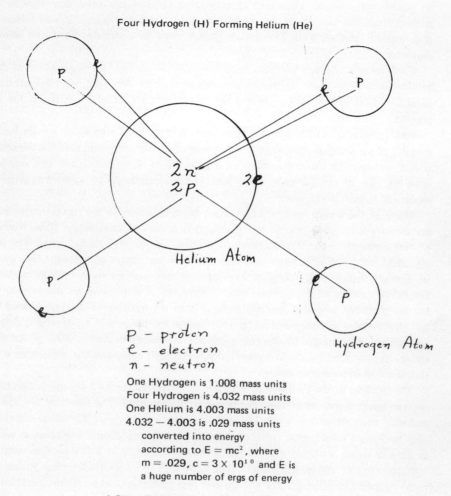

P — proton
e — electron
n — neutron

One Hydrogen is 1.008 mass units
Four Hydrogen is 4.032 mass units
One Helium is 4.003 mass units
4.032 − 4.003 is .029 mass units
    converted into energy
    according to $E = mc^2$, where
    $m = .029$, $c = 3 \times 10^{10}$ and E is
    a huge number of ergs of energy

A Simplified Version of a Nuclear Synthesis

CHAPTER 6

THE PLANETS

Since a star and its attendants make up a solar system the number of candidates for the category is enormous. Even after disqualifying stars for one reason or another the number capable of having a system of encircling non-luminous objects is in the billions. Extending the principle of the uniformity of nature to the solar system entities capable of supporting life, the result is an astronomical figure. Those holding intelligent life would also be very large. Recognizing this deduction, astronomers at an international conference meeting in Armenia, U.S.S.R., in 1971, urged the nations of the earth to plan a program to search for extra-terrestrial life.

The hard evidence for the existence of other solar systems, besides the complex governed by the sun, is very meager. Each suspect star is followed for years to determine whether its tiny motion across our line of sight is influenced by unseen objects. If the star's plotted position over a score or more of years forms a straight line, then the star has not been subject to the gravitational influence of an accompanying planet. When the plotted position forms a wavy line, some factor, a planet or a close star, must account for the motion.

The first two stars suspected, 61 Cygni and Lalande 21185, have accompanying masses about 1/100 of the sun, and the unseen objects may be very small, dim stars or very large planets. The third star evaluated, Barnard's star in the constellation Ophiuchus, has the largest motion across our line of sight; it moves a little more than 10 seconds of arc per year. Peter van de Kamp (1901–   ) at Swarthmore College Observatory in Pennsylvania charted the motion over several decades and inferred the presence of a companion small enough to be classified as a planet. Although the object has a mass 50 percent more than does Jupiter in our own solar system and is 7/100 of our sun's mass, the mass is about 1/100 of that of the parent star, making it the smallest body ever detected beyond the sun's attendants.

Only half-hearted attempts have ever been made to contact intelligent life elsewhere. One was by Frank Drake (1930–   ) and associates at the Green Bank Radio Observatory in West Virginia when during their Project Ozma, the radio telescope was used a few hours per month in an effort to catch radio signals formed by an intelligent being. Another similar exercise was in the U.S.S.R. at the Soviet Radio-Physics Institute in Gorky under the direction of V. S. Troitsky. Suggestions never carried out include the drawing of a huge right triangle on the sands of the Sahara Desert on the theory that intelligent extra-terrestrial life would be acquainted with the Pythagorean doctrine equating the square of the hypotenuse to the sum of the square of the other two sides.

Intelligent life on earth has found out much about the solar system governed by our sun. The information has grown exponentially since the birth of modern science despite the diligent observations of earlier men. More specific details as well as general data about planets are now available.

All the planets are known to be more or less spherically-shaped, largely solid bodies spinning about an axis while going around the sun. As seen from the earth they move in the 18°-wide zodiacal region displaying, like the sun and moon, eastward drift among the stars. Unlike the sun and moon, only planets periodically retrogress or move westward among the stars. Mercury briefly reverses its eastward motion through the stars every 116 days; Venus does the same every 584 days.

All the planets confirm to Kepler's laws of planetary motion. The first law is a modification of the contention of Copernicus (1473–1543) that the orbits are circular. Kepler (1571–1630) said that the planets move in elliptical orbits about the sun, with the sun being in one foci of the ellipse. The orbits are not too eccentric but they are definitely elliptical. Kepler's second law describes the speed of revolution through the statement that the radius vector sweeps out equal areas in equal times. In other words, a planet does not cover the same orbital distance in a given time but the areas carved out by a line from the sun or a line from the other focus to the position of the planet are equal. An inference from the third or harmonic law also supports the view that the planet travels at a greater speed when nearest the sun. Kepler stated the third proposition after ten years of effort searching for the relationship. He said that the squares of the periods of any two planets are proportional to the cubes of their distance from the sun. Today, the expression can be paraphrased as the ratio of the square of the time a planet takes to go about the sun divided by the cube of its average distance to the sun is a constant equal to the sum of the masses of the sun and planet; or T^2/D^3 is a constant.

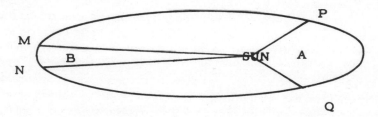

Area A = Area B
when
time to travel arc MN = time to travel arc QP

Kepler's Second Law: Area A = Area B (Ellipse Exaggerated)

Kepler, the former theology student, believed god to be a geometer, displaying the five regular solids, those with equal faces, within the solar system. Kepler claimed that if a cube were inscribed in a sphere containing the orbit of Saturn, then thought to be the outermost planet, the sphere holding the orbit of Jupiter would just fit within the cube. A tetrahedron, having four equilateral triangles as faces, as well as the other regular solids such as dodecahedron and icosahedron would fit into the remaining planetary orbits. Kepler was loathe to relinquish this concept even later in life, despite the inability of the facts to support the idea.

In Kepler's time there were no general rules for the distances of the planets from the sun. Late in the eighteenth century, a generalization was devised that seemed to work. The formula called the Titius-Bode law, or Bode's law, is best presented by first writing a row of 4s. Under the first 4, 0 is placed; then 3 under the next, twice 3 or 6 under the third, twice 6 or 12 under the fourth and so on. The sum of each column divided by 10 yields the relative distance of the planets, where the earth's distance is one.

.4	.4	.4	.4
0	.3	.6	1.2
.4	.7	1.0	1.6

Bode's Law for the First Four Planets

The discovery of Neptune and Pluto led to the abandonment of the rule since they did not conform. A series of fractions and whole numbers 1/3, 2/3, 1, 1.5, 5, 10, 20, 30, 40, now best represents the planetary distances compared to the earth's average distance. Accordingly, the planet Pluto is about 40 times the earth's 93,000,000 miles from the sun.

The planets are in practically the same slice of space. They encircle the sun in almost the same plane, making elliptical orbits. Each planetary orbit has a unique eccentricity and a unique time of revolution.

The sidereal period of a planet is its time of revolution around the sun, from a star to the same star again, as seen from the sun. By comparison to the moon's sidereal period, a planet's is likewise the interval between two successive times a straight line can be drawn between sun, planet and distant star. Planets, like the moon, also have a synodic period and again it is the interval between two successive times a straight line can be drawn between sun, planet and earth.

Other terms useful for the moon applicable to the planets are elongation, conjunction and opposition. As in the moon's case, the words refer to the arc distance as seen on the sky of the earth. Elongation is the most general description of how far the planet is from the sun; it can be as much as 180°. A planet is in conjunction when the earth, sun and planet lie most nearly in a straight line with the earth at one extremity. This elongation of 0° occurs when the sun and planet are seen in the same direction. A planet is in opposition when earth, sun and planet are again most nearly in a straight line with the earth in the middle. This elongation of 180° occurs when the sun is seen in one direction and the planet is 180° away.

Earth   Sun   Planet   Sun   Earth   Planet
CONJUNCTION   and   *OPPOSITION*

All planets have elongation and can be spotted at the conjunction or opposition point. Yet years before instrumental analysis of planets began, the first two planets, those between the sun and earth, Mercury and Venus, were called inferior planets while the others came to be known as superior planets. The terms are not pejorative, as study of the planets shows.

Mercury is a difficult planet to see with the unaided eye because of its closeness to the sun. The planet never strays more than $28°$ from the parent star and consequently observers try to capture a view at dawn or dusk. Early peoples called the morning and evening views by different names. The Greeks used Mercury for the evening one and Apollo for the other; the Egyptians had the names Horus and Set, respectively, while the Hindu corresponding titles were Raulineya and Buddha.

Courtesy: NASA

Mercury, photographed by Mariner 10, March 29, 1974, at a distance of 124,000 miles.

Telescopic sightings avoid the long path through the earth's atmosphere at dusk or dawn. Professional astronomers study the planet when it is highest in our sky, during the day.

Because the planet is the closest to the sun several other pieces of information are immediately apparent. For one, Mercury receives the most light and heat. The contention that it is the warmest planet is not tenable because heat absorbed and retained depends upon the composition of planetary material; the rug and tile on a bathroom floor receive the same amount of heat yet one is much colder than the other. Then in accordance with Kepler's second and third laws of planetary motion, Merucry has the fastest motion about the sun. One revolution about the sun takes 88 days.

Another superlative about Mercury is that it is the smallest planet with a diameter of little more than 3,000 miles. It is the least massive, as determined by its gravitational effects on Venus space probes such as Mariner II and Mariner V. The passage of the asteroid Icarus 10 million miles away in April 1968 also helped determine Mercury's mass to be about 1/18 of the earth's mass. It reflects only about 6 percent of the sun's light incident upon it and this is less than any other planet's albedo.

Some other characteristics are almost superlative. Its orbit is highly eccentric, the planet being 28.6 million miles away when close to the sun and 43.4 million miles away when farthest. Only Pluto has a more elliptical path about the sun. The orbit of Mercury has a $7°$ inclination to the central plane of the sun, and again only Pluto has a more inclined orbit. Its average density is close to that of the earth's, 5.5 times that of water. Mercury has no satellites but such is also the case for Venus and Pluto. Mercury's tiny atmosphere was detected in 1974, while Pluto probably has no atmosphere.

Mercury is the only planet involved in a substantiation of the theory of general relativity. Along with other planets, its orbit is slowly rotating; a line through Mercury and the sun is moving 574″ every 100 years. The amount calculated to occur, because of gravitational action by other planets and solar system objects, is 531″. The difference between the observed and calculated, 43″, was at one time assigned to the presence of an undetected planet. Named Vulcan and ardently sought in prior centuries, the undetected planet as an explanation was discarded with the advent of relativity theory. It predicted the observed value of 574″ per century.

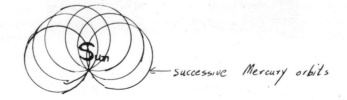

— *successive Mercury orbits*

Exaggerated Version of the Slow Advance of the Perihelion of Mercury, 574″ per Century

Venus is the unique because it is the brightest planet, is the most perfect sphere and has the most nearly circular orbit among the nine major parasites of the sun. Almost every other distinctive feature is matched by some other planet.

Venus is almost the same size as the earth; the diameter is only about 400 miles less than the earth's. The planet's density is like that of Mercury and the earth, being about 5.2 times that of water. Like Mercury, Venus is sometimes seen in the morning and sometimes in the evening; the early Greeks called the morning object Phosphorus and the evening one, Hesperus.

Being an inferior planet, like Mercury, Venus shows phases. Galileo (1564–1642), using his primitive sky glass, was the first to call attention to the fact. Those who supported the earth-centered view found the phenomenon difficult to explain. The Ptolemaic conception then in vogue held that a straight line could always be imagined between the sun, Mercury, Venus and the earth. Such a qualification would prevent the formation of an alignment wherein earth observers could see Venus in portions of a full circle.

Venus never strays more than 47° away from the sun. Since diurnal motion is 15° an hour, Venus is at most about three hours behind the sun in setting in the west. Consequently earth observers, except those in polar regions, will never see Venus after midnight.

Venus may be a candidate for having an equality of rotation and revolution periods. The planet moves with an average speed of 22 miles a second to go about the sun in 225 days. Its rotation period, determined by radar, is about 250 days. Should the rotation and revolution periods be equal, as was thought to be true for Mercury prior to 1965, then only half of the surface of the planet is sunned; earth observers would only see 50 percent of the surface because the same portion of the planet faces the sun continuously.

The radar observations made in the early 1960's did show Venus to rotate from east to west. Every other planet, except Uranus, spins in the other direction, from west to east.

Venera 4, the Soviet probe passing by Venus on October 18, 1967 had instruments that made a partial chemical analysis of the Venusian atmosphere. The result was 90 to 95 percent carbon dioxide, one percent water vapor and .4 to .8 percent oxygen. Both Venera 4 and the American Mariner V, passing about 250 miles from the planet's surface on October 19, 1967 showed the Venusian atmosphere to be very dense, at least 20 times that of the earth's atmosphere at sea level.

Radio waves penetrating the optically opaque atmosphere of Venus revealed an area 910 miles across (the size of Alaska) near the equator with a dozen large craters ranging from 21 to 100 miles in diameter. All the craters are very shallow, the largest being only one-quarter of a mile deep.

Mars is unique because it holds the two smallest natural moons in the solar system. Called Phobos and Deimos, meaning "fear" and "panic" respectively, they were discovered in 1877 by Asaph Hall (1829–1907) at the United States Naval Observatory. However, Johannes Kepler (1571–1630) earlier had argued that Mars should have two moons. In 1726, Jonathan Swift (1667–1745) in his *Gulliver's Travels* described such moons and so did Voltaire (1694–1778).

The photographs taken by Mariner 9 were very revealing, indicating an earth-like planet. Very large volcanoes, huge faults, rift valleys and channels are present. Two types of the latter appear to be water-eroded tributaries and valleys.

Courtesy: NASA

Venus, photographed by Mariner 10, February 6, 1974, from 450,000 miles.

Only Mars, among all the planets, is considered by some to be, like earth, an abode of life. The evidence for the belief is flimsy but so is evidence for the contrary proposition. Advocates of the life on Mars cite an array of facts such as the periodic change of surface color, the periodic change in the size of the polar caps, the almost similar rotation period and inclination as the earth, the presence of water vapor and a trace of oxygen in the atmosphere and the citing by some of an array of seemingly straight lines on the surface. Those who deny the existence of life on Mars point to our failure to detect chlorophyll, a necessary chemical for photosynthesis by green plants, the tiny amounts of oxygen and water vapor, the lower temperatures than on earth, and the 20 photographs of the surface taken by Mariner IV in July, 1965 did not reveal any so-called lines called canals.

The question of life on Mars should instigate discussion about the definition of life. Should a classical biological view be taken then evidence of metabolism, growth, reproduction and irritability must be shown in at least one Martian organism. If life is to be seen centered about the DNA molecule, then such a chemical must be detected. Those who would have no real distinction between life and lifeless, viewing the former as a more complex extension of the latter, would have no real problem.

The contention that life on Mars in no way resembles the phenomenon on earth and a "different form of life" exists is at variance with the uniformity of nature principle. The essential features must be the same on both planets, with variations being as wide as they are on earth. Those who press for a "different form of life" on Mars need to outline the differences and show how to detect the presence of the forms.

Investigators on earth enchanted with the theme of life on Mars have reconstructed a Martian earth in which unspecialized earth organisms can live. Algae and fungi, plants without roots, stems, leaves, and flowers, placed in what could be Martian soil did not die. Such positive evidence, however, is often mixed with imagination such as the contention that the moons of Mars, being so small, are really artificial satellites sent up by an ancient Martian civilization. The speculation is then bolstered with more imagination by the statement citing how the light reflectivity of the moons resemblances that of polished metal.

Telescopic observation and other instrumental analysis of Mars and its satellites from the earth or near the earth will eventually answer some of the questions about life on Mars. Such study is best at intervals of the synodic period of Mars, every 780 days, when Mars is at opposition. The earth is then closest, but not all oppositions of Mars are equally favorable. The distance from the earth can be anywhere from 35 to 63 million miles due to the eccentricity of the orbit. The best time is when Mars is at its perihelion, its closest approach to the sun and this occurs once or twice every 15 to 17 years. In 1971 the distance of earth to Mars was the shortest, 35 million miles. Three probes, two Russian and one American swept by Mars, November, 1971, and measured some values. The entire surface was photographed, and arguments ensued about the possibility of some features being caused by water erosion.

The gravitational attraction between Mars and an interplanetary probe indicates the mass of the planet is only .107 that of the earth, although half the size with a diameter of about 4200 miles. Because the Martian year is longer than the earth's, the seasons on Mars are twice as long; its southern hemisphere climate is more accentuated with the polar cap reaching near the equator in winter and veritably disappearing in summer. Mars is extensively cratered, resembling the surface of the earth's moon. The fact that the southern hemisphere of Mars is extensively cratered, and not the northern half, is perplexing. So are its channels, the nature and composition of the polar caps and the volcanic activity.

Jupiter is a unique planet because it is the largest, with a diameter of about 87,000 miles, the fastest spinning, with a rotation time of nine hours and fifty minutes, the one with most satellites, thirteen, and one with a giant red spot, 30,000 miles across. One of the oldest theories to account for the spot maintains that some kind of solid substance is floating in the planet's dense lower atmosphere, revealing its upper surface above the clouds. Yet no known substance would be able to float in the hydrogen-helium atmosphere. The Taylor column model is the contention that the spot is caused by either a bump or depression on the planet's surface. As the planet's atmosphere flows around the feature, a tube of stagnant gas called a Taylor column forms. Both the raft and Taylor column theories are combined in the suggestion that a light layer of solid hydrogen is floating in a hydrogen-helium atmosphere. The base of the layer is heated and rises; some hydrogen melts, heat is released and the layer sinks to a lower level where the cycle again begins. The spot could also be a huge hurricane.

Jupiter is unique because it has an abundance of hydrogen, making it a candidate for once being a star. One estimate is that the planet is 78 percent hydrogen; helium is also abundant while methane and ammonia, easily detected in Jupiter's atmosphere, make up about one percent of the atmosphere; in 1974, ethane and acetylene were also found.

Jupiter, like the other planets, is unique because of its physical and orbital characteristics. A glance at the table of properties of planets reveals that no two have identical numbers for important properties. Not shown in such tables, however, is another individuality of Jupiter — the vast number of charged atomic particles circulating around the planet. The interpretation of radio energy from Jupiter prompts the conclusion. The planet also emits more heat than it receives from the sun.

The most special thing about Saturn is its system of encircling rings. When Galileo (1564–1642) saw them his view was such that he thought he saw four satellites; about 45 years later in 1655 Christian Huygens (1629–1695) made the interpretation accepted today. The entire ring system has a diameter of 276,000 kilometers (171,000 miles) and is composed of an outer ring, a bright ring, a semitransparent gauze or "crepe" ring, and a very faint inner ring. The last, uncovered in 1969, spans the distance between the planet and the crepe ring. The latter is about 11,500 miles wide, the bright one is 16,000 miles wide and the exterior ring is about 10,000 miles wide. The entire system is very thin with a thickness anywhere from a kilometer to ten miles at the most. During the middle of the nineteenth century the British physicist James Clerk Maxwell (1831–1879) showed that the rings must be composed of small objects; about a century later investigation centered on whether the moonlets were composed of ice or paraformaldehyde.

Another unique characteristic of the giant planet is its average density. The weight per unit volume is so small, .69 grams per cubic centimeter, that the entire planet could be floated in water.

Saturn, like Jupiter, has a yellowish cast and bands of encircling color. Near the poles blues and greens predominate while those near equatorial regions are darker. Small yellowish-white spots have been seen; one observed in 1933 encircled three-fifths of the circumference and then disappeared.

Uranus has several elements of distinction as a planet. For one, it is the first to be discovered; on March 13, 1781, William Herschel (1738–1822) found it and was dissuaded from calling the object King George's star. At first he thought he had a new star rather than a planet. The astronomer royal of England, J. E. Bode

(1747–1826) suggested the name Uranus. Herschel reputedly fled to England to escape military service. For 16 years he was an organist at the Octagon Chapel in Bath. He took up optics and astronomy and "resolved to take nothing upon trust but to see with my own eyes all that other men had seen before". He came to the conclusion that "seeing is an art which must be learned." When he was 36 years old, after plenty of failures, he constructed his first reflecting telescope. In time, he became an adept builder of telescopes and earned substantial funds at the task. He was already well-to-do when he married a wealthy widow.

Uranus is unique because the inclination of its equator to its orbit is 98°. The planet has "fallen flat on its face". With such an alignment, the poles of Uranus periodically face the sun.

The planet appears as a greenish disk, brighter at the center than at the limb. Uranus also has belts of color but much less pronounced than these of Saturn and Jupiter.

Irregularities in the orbit of Uranus led to the discovery of Neptune in 1846. The actual event was a comedy of errors wherein established astronomers failed to cooperate with younger ones eager to spot the planet. In England, the astronomer royal did not help while in France, a young scientist had to seek help from a German observatory.

Neptune appears green as does Uranus, has two moons as does Mars, has an almost circular orbit as does Venus, contains methane and hydrogen in its atmosphere as does Jupiter, and is closest in diameter to Uranus. The latter has a diameter of 47,000 kilometers while Neptune's is 45,000 kilometers. Since Neptune takes 165 years to make one revolution about the sun, its first complete revolution since discovery by man will occur in 2011.

Pluto, the outermost planet was also predicted long before its discovery. Percival Lowell (1855–1916) in 1915 made a very accurate description of the eccentricity, and period of revolution. Not until 1930, did amateur astronomer Clyde Tombaugh (1906–   ) first detect the planet.

Among the unique properties of Pluto, besides its greatest eccentricity and orbit inclination of all planets is the difficulty of measuring its mass, size and density; there are no definite figures. Then, too, Pluto probably began as a satellite of Neptune; between 1979 and 1999, Pluto will be inside Neptune's orbit and the latter will be the outermost planet.

Planets beyond Pluto have not been detected. If any is present, some astronomers have already chosen a name for the next: Proserpine.

SELECTED REFERENCES

Ray Bradbury et al, *Mars and the Mind of Man*, New York: Harper and Row, 1973.
Morton Grosser, *The Discovery of Neptune*, Cambridge, Mass.: Harvard University Press, 1962.

The Planets

	Diameter km.	Diameter Earth=1	Density	Rotation	Inclination of Equator to Orbit	Oblateness	Gravity at Surface Earth=1	Albedo
Mercury	4,880	.38	5.1	59d	?	0	.39	.06
Venus	12,112	.95	5.3	242.9d	23°	0	.91	.76
Earth	12,742	1.00	5.52	$23^h56^m04^s$	23°27'	1/297	1.00	.39
Mars	6,800	.53	3.94	$24^h37^m23^s$	24°	1/192	.38	.15
Jupiter	143,000	11.19	1.33	9^h50^m	3°	1/15	2.64	.51
Saturn	121,000	9.47	.69	10^h14^m	27°	1/9.5	1.13	.50
Uranus	47,000	3.69	1.56	10^h45^m	98°	1/14	1.07	.66
Neptune	45,000	3.50	2.27	16^h	29°	1/40	1.41	.62
Pluto	<6,000	<.47	?	6.387d		0	?	?

	Sidereal Period	Synodic Period	Inclination Orbit	Orbit Eccentric	Distance Sun Earth=1
Mercury	87.97d	115.88d	7°.004	.20563	.3871
Venus	224.70d	583.92d	3°.394	.00679	.7233
Earth	365.26d			.01673	1.000
Mars	686.98d	779.94d	1°.850	.09337	1.5237
Jupiter	11.86y	398.88d	1°.305	.04844	5.2028
Saturn	29.45y	378.09d	2°.490	.05568	9.5388
Uranus	84.01y	369.66d	.773	.04721	19.182
Neptune	164.79y	367.48d	1°.774	.00858	30.058
Pluto	247.68y	366.72d	17°.170	.25024	39.439

CHAPTER 7

OTHER SOLAR SYSTEM OBJECTS

Besides the sun and planets, the solar system contains thousands of smaller planets, 33 natural satellites, an unknown number of comets, a very large number of meteors, and a growing amount of objects sent into space by men. Material called tektites, found on the earth, may also be classified as part of the solar system.

The name asteroid, starlike, was suggested by William Herschel (1738–1822) for small planets. In some sections of the United States the term planetoid is employed. Astronomers also refer to "vermin of the sky" and occasionally the descriptive phrase "flying mountains" has been employed for the minor planets.

The first asteroid, Ceres, to be detected was by the Sicilian astronomer, G. Piazzi (1746–1826) in 1801. He chose the name of the tutelary divinity of the island also known as the goddess of grain. Bode's law for relative planetary distances had predicted the existence of an object where Ceres was seen. Since Piazzi's time, hundreds of asteroids have been spotted in the region between Mars and Jupiter. In one search for satellites of Jupiter, 32 new asteroids were found on the photographic plates. Practically all the asteroids are under 100 miles in diameter. Ceres is unusual with a 770-mile diameter; both Pallas and Vesta are about 480 miles wide and all the others are much smaller with the large majority being under 100 miles wide. The sizes of the smallest are estimated from brightness measurements, with the assumption that they reflect as much light per square mile as do the largest asteroids.

The very eccentric orbits of the small planets occasionally brings them close to the earth. Hermes, spotted in 1937, one mile in diameter with a mass of three billion tons, came within 485,000 miles of the earth on October 30, 1937. Betulia, one-half a mile in diameter, discovered in 1950, was 14,600,000 miles away from the earth in May, 1963. Amor came within 16,200,000 miles from the earth in January, 1931 and the English astronomer Sir Harold Spencer Jones (1890–1960) used it to determine more exactly the distance of earth to sun; he found 93,009,000 miles and the value was superseded later by using transits of and radar observations of Venus.

Icarus has an orbit bringing it very close to the sun, coming within 17 million miles; at the other side of its orbit, the microplanet is 183 million miles from the sun. Hidalgo has the most unusual orbit, reaching out to within Saturn, but inclined 43° from the common plane of the planets.

The 1969 edition of the U.S.S.R.'s *Minor Planet Ephemeris* listed data for 1,735 asteroids, and since then other interesting ones have been uncovered such as Geographos, coming within six million miles of the earth in 1969. The book does contain information about the Trojan asteroids predicted in 1772 by Joseph Louis

Lagrange (1736–1813). He showed that there were two points in the orbit of Jupiter where a microplanet could be possible; where Jupiter, the sun and the objects could form equilateral triangles, all sides equal. About 150 years later the objects called Trojans were found. Generally, those to the east of Jupiter have been named after the Greek Homeric heroes while those to the west have Trojan names; before this convention was adopted, a Greek name slipped in among the Trojans and a Trojan name among the Greeks.

Altogether fourteen are known.

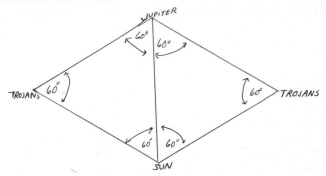

Position of the Trojan Asteroids

The origin of the asteroids as a whole can of course be assigned to one of three theories: They could at the birth of the solar system have been formed in the manner of their current existence; a planet could have broken up to make them; they may be a planet in the making. Any of these must encompass the fact that the combined mass of all the asteroids is probably less than a thousandth of the mass of the earth. Collision between two bodies is an independent idea for which there is some support. The asteroids can be grouped according to similar orbital characteristics and two subgroups encompass a large number of the objects.

Some other solar system material may have once been asteroids. Jupiter's outer satellites, for example, are small unlike the other moons; so are those of Mars. Perhaps the outer satellites of Jupiter were once Trojan asteroids that came too close and were captured.

The 33 natural satellites of the solar system are not equally distributed among the planets. Mercury, Venus and Pluto have none. During the seventeenth century, however, several competent observers described a satellite of Venus. Jean Dominique Cassini (1625–1712), discoverer of four satellites of Saturn and the first of several generations of his family to become astronomers, was one; Johann H. Lambert (1728–1777) even published in 1773 a treatise on the moon of Venus; Frederick the Great wanted to honor Jean Le Rond d'Alembert (1717–1783) by naming the moon after him but the latter refused the honor. The astronomers who saw a moon of Venus had either seen nearby stars or ghost images produced by their telescopes.

The two moons of Mars are mentioned in the prior chapter. They are the smallest in the solar system and, unlike the tiny ones of Jupiter, are fairly close to the main body. The inner one, Phobos, is about 5800 miles distant while Deimos is 14,600 miles from Mars.

The first four of Jupiter's thirteen, uncovered by Galileo (1564–1642) in 1610, have names and are sizable. Their radii are 1,775 for Io, 1,550 for Europa, 2,800 is Ganymede's and 2,525 kilometers is Callisto's while all the others have radii under 50 kilometers. Moreover, some of the outer satellites have retrograde revolution and have eccentric orbits highly inclined to the plane of the planet's equator. Ganymede's atmosphere was first detected in 1973; another natural satellite with an atmosphere encircles Saturn.

Only one of Saturn's ten satellites is comparable in size to Jupiter's largest. Titan is distinctive because it is a natural satellite having an atmosphere, probably methane. It also has a reddish cast. Saturn holds a recent one to be detected, Janus, in 1966.

The five moons of Uranus move in a retrograde fashion in a plane almost perpendicular to that of the parent planet's orbit. The large axis inclination of Uranus is the reason. A Bode-type law has been formulated for the distance from the planet for the moons of Uranus as well as the first five of Jupiter and the first five of Saturn. However, the constant and the number to be added is different for each planet.

The two satellites of Neptune are unlike. Triton is large with a circular orbit and has retrograde revolution. Nereid, found in 1949, is small and moves in the same direction as does the planet but in a highly eccentric orbit.

The moon of the earth is the only natural satellite visible to the unaided eye. Vesta is the only asteroid that can be seen without a telescope. There are many more comets in this category; 400 were bright enough to have been recorded before the introduction of optical devices. In recent years several have appeared. One bright enough to be seen during the day was observed in 1910. Comet Arend-Roland and Comet Mrkos were naked-eye comets in 1957. In 1965 comet Ikeya-Seki was a daylight phenomenon.

In accordance with the prevalence of astrology and the awesome appearance of some comets, early peoples believed comets to be omens of evil. According to William Shakespeare, "When beggars die there are no comets seen; the heavens themselves blaze forth the death of princes". At the beginning of the eighth century, Bede, the English monk who wrote an ecclesiastical history of England, commented, "Comets portend revolutions of kingdoms, pestilence, war, winds and heat". Presumably our age produces more sophisticated people but in 1910 a man became wealthy selling comet pills.

About two thousand years ago, Seneca suggested that comets were celestial bodies which could recur periodically, but the scientific study of them begins with Tycho Brahe (1546–1601). He showed the comet of 1577 to be more distant than the moon and traveled about the sun. Isaac Newton (1642–1727) demonstrated how the comet of 1680 moved, in accordance with the law of gravitation, in an elliptical orbit of great eccentricity. Edmond Halley (1656–1742) noticed a resemblance among the orbits of the comets of 1531, 1607, and 1682 and deduced them to be the same body that would return in 1758; it came again in 1835 and 1910 and should reappear in 1986. As of the beginning of the last third of the twentieth century, 56 comets have been observed more than once.

A comet first appears as a star-like object surrounded by a luminous fog and an attached large stream of hazy light. The comet may have only a suggestion of a tail, many tails or irregularities in the tail but generally the tails point away from the sun. There are great variations in size, shape, tail structure, brightness and composition. Many different chemical elements and some combinations of them have been identified in the head and tail of comets.

Among the comets that have been studied is Encke's, named after J. F. Encke (1791–1865) who investigated it for several revolutions after its first citation in 1786. It has the shortest known period, 3.3 years. He found the object came back 2 hours earlier than predicted, and since 1860 it has been arriving an hour earlier on each successive revolution. Perhaps material is being vaporized away with passage near the sun.

The "gravel-bank" model for a comet, a popular theory, holds the nucleus to consist of fine gravel and dust with gases on the surface of the solid particles. The idea would account for the gases appearing to stream into the tail. A competing conception describes the nuclei as fairly compact solid bodies composed of frozen gases, such as water, ammonia, and methane together with dust and parts of molecules called free radicals. The dirty snowball idea would better explain why comets coming remarkably close to the sun are not destroyed and the persistence of some of the periodic comets. It could also help to account for the wayward motion of comets on the grounds that evaporation could produce a reaction contrary to the forward travel.

Another widely-held idea is that billions of comets hibernate in the region between the sun and stars. The random passage of a star disturbs the motion of some comets enough to make them move into the gravitational field of Jupiter, Saturn or a planet at the outskirts of the solar system.

Other suppositions include the idea that asteroids are spent comets and the direct impact of a comet with the earth was the cause of the disaster in Siberia, June 30, 1908. The catastrophe in the Tungus forest leveled many square miles of trees and knocked down people more than 30 miles away. The pattern of the damage indicated an explosion hundreds of feet above the ground.

More factual is the association of comets with meteor showers. Biela's comet was last seen in 1852 but when the earth crossed the orbit of the comet in 1872, a giant-size meteor display occurred on earth. At one location in Italy over 30,000 meteors were counted within 6.5 hours. The Perseids, the stream of meteors emanating from the constellation Perseus in Mid-August have been linked with the debris left behind by Swift's comet also called 1862 III, the third comet to pass near the earth that year. The Perseids are known, too, as the tears of St. Lawrence, after the saint who suffered martyrdom by being roasted to death in Rome on August 10, 158 A.D.

Meteors, also known as shooting stars, can be seen as bright streaks of light. They are bits of stone and metal heated to incandescence by passing through the earth's atmosphere at speeds from 7.5 to 50 miles per hour. Most are burned up before reaching the surface and come to the ground as dust. Perhaps two million tons a year are thus added; all since the formation of the planet would make a layer about ten feet thick.

Meteor showers, emanating from a different constellation each time occur every month except February, June and September. On the other hand, individual meteors can be seen every clear night, preferably after midnight. If exceptionally bright the phenomenon is called a fireball; an exploding fireball is called a bolide.

When a meteor falls to the surface of the earth as a solid body it is called a meteorite. The oldest recorded fall for which a specimen exists landed at 11:30 A.M. on November 16, 1492. It is now in a locked glass case in the town hall of Ensisheim, Alsace, France, near where it fell. However, a stone venerated by the Mohammedans is believed to be a meteorite. It is called the Right Hand of God on Earth and is said to have been dropped from Paradise when Adam was created. Every devout Moslem turns to it when he prays. The dark reddish-brown stone is built into the northeast corner of the Kaaba in Mecca.

The largest meteorite on earth, about a fifty ton mass, is near Grootfontein, South West Africa. The largest in captivity weighs 34 tons and is at the American Museum of Natural History in New York City; it was found by Admiral Peary in Greenland in 1897.

The Port Orford, Oregon meteorite was found in 1859; specimens were sent to several museums. Plans to ship the entire object never materialized because the meteorite seemingly disappeared. One found near Willamette, Oregon in 1902 weighing 14 tons was hauled away by the discoverer who charged admission to see the object. The owners of the land on which it was found, the Oregon Iron and Steel Company, sued and recovered the meteorite.

The Siberian fireball of February 12, 1947, near Vladivostok, spewed many fragments over an area of two square miles. Trees were felled; in other falls animals were killed. Yet there is practically no substantiated cases of a meteorite striking a human being. In 1954 a woman in Alabama was supposed to have been struck and one in Persia reputedly killed a person. Cases which have been checked involve meteorites striking an automobile in a garage or an unoccupied outhouse. The infrequency of human contact seems likely in the light of the tiny area offered by three billion persons and the much larger area of the earth.

On August 10, 1972, a meteor with a diameter of about 13 feet sped along a course in the earth's atmosphere, and was tracked from Salt Lake City, Utah to Calgary in Alberta, Canada. It reached a height of about 36 miles above the surface in eastern Idaho. It zoomed out of the earth's atmosphere without breaking.

Fortunately, such meteors, having tremendous potential for destruction and capable of escaping from the earth although already within the earth's field of attraction, are exceedingly rare.

During the 1960's a fierce scientific argument ensued over the presence of "formal organisms" in certain types of meteorites. Analysis of one that fell in France in 1864, the Orgueil meteorite, showed organisms to some scientists and no such thing to equally-competent investigators. Other portions of the same meteorite were analyzed and contaminants such as weeds, evidently planted as a hoax, were found. Nonetheless the argument continued as other types of the same kind of meteorite were checked. More fuel was added to the fire by the detection of carbon-containing compounds such as formaldehyde, in the space between the stars. Lines seem to be firmly drawn between investigators who believe meteorites contain life forms and those who do not.

The type of meteorite said to contain life forms are carbonaceous chondrites, a division of the most common kind, stony or aerolites, resembling rocks on earth. The second type, irons or stony irons, siderites and siderolites, when cut, polished and etched generally show a characteristic crystalline structure called Widmanstätten figures.

Metallic meteorites are responsible for forming craters of all sizes. The first crater to be discovered on earth, and the most commercialized, near Winslow, Arizona, is about 4,200 feet across and 600 feet deep. The New Quebec Crater in Quebec discovered in 1950 via aerial photographs, is twice as large and is now filled with water. Another huge crater is in Algeria but as in the Canadian case, no meteorites nor fragments of such have been found in the area.

Meteorites may not be the only solid material on earth of astronomical origin. Tektites are glassy objects the size of gravel found in basically four places. The largest and youngest tektites are in the southwest Pacific area — the southern part of Australia, Indochina, China, the Philippine Islands and Indonesia. The next older tektites, dated by contained radioactive materials, are found in the Ivory Coast Republic. The Czechoslovakian tektites named after the Moldau River, Moldavites are not dark brown as are the first two, but are light and olive green and occasionally brownish. The oldest tektites are in America, in east-central Texas and south-central Georgia.

Tektite glass has a tiny amount of water, less than man-made glass and much less than the volcanic material called natural glass or obsidian. Some tektites have bubbles with a more or less high vacuum.

The earliest mention of tektites described them as a special kind of volcanic glass. Other hypotheses later considered include prehistoric manufactured glass, sand or soil melted and fused by lightning, and the result of meteorite bombardment. Those supporting the last view claim molten material was splashed high in the atmosphere on a ballistic trajectory by a huge meteorite striking the earth. Thus the Czech tektites were created by a meteorite that produced the 16-mile crater called the Ries Kessel (Giant Kettle) near Nordlingen in Southern Germany. Another crater, now a lake in Ghana, is connected with the Ivory Coast tektites. Craters to account for the Pacific and American tektites have not yet been found.

A competing theory holds that the tektites were formed by an impact on the moon and droplets splashed off to the earth. A reason for the distribution pattern on earth has yet to be formulated. Of course, it is possible to claim that man transported the tektites from one or two places; a single tektite was found on Martha's Vineyard in Massachusetts, identical to the Georgia ones, but others were not detected despite a diligent search.

SELECTED REFERENCES

Zdeněk Kopal, *The Solar System*, New York: Oxford University Press, 1973.
Patrick Moore, *The New Guide to the Planets*, New York: Norton, 1972.

CHAPTER 8

DEVELOPMENT OF THE SOLAR SYSTEM

The existence of any natural objects or group of objects prompts not only a sense of wonder but also the query of how it came to be what it is. The last question is much different than asking about its creation, an inquiry suitable for theologians and philosophers but not for scientists. They are concerned with changes occurring with time, development, and not the making of something from nothing.

Well-accepted theories of development in natural science include the doctrine of organic evolution. Astronomy, however, does not have any development idea in such a favorable category. Indeed, the conceptions about the development of the solar system have been called speculative. Nonetheless, the ideas serve a purpose, displaying the fact that science deals with information of all degrees of certainty from established fact through educated guess; moreover, much can be codified with a theory, even one eventually a failure.

Conceptions about the development of the solar system fall into two broad categories, nebular and dynamic encounter theories. The latter envision a collision or a gravitational attraction between two or more objects. During the middle of the twentieth century dynamic encounter theories became less popular because astronomical objects that could lead to solar systems were considered to be in some regular array with little chance for approaching one another. The billions of stars in our galaxy are in such locked situations while impact or attraction between less massive bodies such as comets and asteroids would probably not lead to solar systems. Nonetheless, proponents pointed to a star that with our sun was the progenitor of our present planetary configuration; a few wondered about the possibility of a primitive Jupiter being the second star involved.

The earliest adherents to either the dynamic encounter or nebular views were philosophers more than scientists. At the end of the eighteenth century naturalist Comte de Buffon (1707–1788) wrote about a collision between a comet with the sun while at about the same time philosopher Immanuel Kant (1724–1804) described a huge hot gaseous globe as the origin of the solar system. Kant could also be viewed as a kind of scientist because according to some authorities, lack of research facilities kept him from the discipline. Another philosopher scientist of the same period, but much more the scientist who proposed a nebular view was Emanuel Swedenborg (1688–1772). The proponent of the view in France was Pierre Simone, Marquis de La Place (1749–1827) and his description has been most publicised despite its being but a footnote in the appendix of one of his books.

According to La Place the present solar system was originally a nebula, a large, hot, spinning sphere of gas. Nebula were then being detected and the supposition of La Place and the others gave nebula a meaningful place in the scheme of things. The theorists could just as well have started with a huge star, or even dancing mermaids. The beginning assumption of a scientific theory is of no consequence towards its acceptance; in some branches of science, the first premise has been contrary to ordinary common sense. The power of the final idea to collate, describe, explain and predict is what counts. Scientists are in this sense pragmatists accepting or rejecting in terms of what the theory can accomplish.

In accordance with Newton's law of universal gravitation, parts of the primeval nebula attracted each other as the whole thing also cooled. The nebula became smaller and consequently its rotation speed increased; just as a ballerina turns slowly with her arms outstretched but more rapidly when held close to the body.

On five different occasions, the nebula was spinning fast enough for the escape of some material at opposite sides of an equatorial bulge. The jets of gas come off in pairs so that five omissions could account for nine planets and the belt of asteroids.

Each ejected mass served as a new nebula to repeat the process for the formation of 33 natural satellites. The original nebula became our sun but the secondary nebulae solidified as did their escaped material to change into planets and moons.

A specious examination of such a nebular theory is appealing. The planets are practically in the same plane of space indicating the common origin from the nebula's equator; all the planets rotate and revolve, securing these motions from the parent body; the same kinds of chemical elements, although in somewhat varying abundances, are everywhere in the solar system; the sun's mass is about 700 times that of all its parasites combined which is likely if the sun is the original material.

Only a modicum of criticism reveals gaping holes in the idea. For one, the sun is not spinning as rapidly as the hypothesis would have; with more than 99 percent of the mass, it should have more than its two percent angular momentum, the product of mass and rotation speed. Another failure of the theory is the inability to account for Phobos, the inner moon of Mars revolving three times as rapidly as Mars rotates. How can a tiny moon turn about a parent body once in eight hours while the planet rotates every 24 hours? The single moon of the Earth and the five of Uranus are problems if satellites were formed in pairs. Another vexing situation is the movement of some satellites opposite to the direction of their planet's rotation and revolution. A cause must also be found for the extreme inclination of the axis of Uranus. Another damaging fact is the alignment of densities, with the innermost planets having an average density much like the earth's while the outermost ones are closer to the sun's average density. If gravitational attraction of the sun contributed to the densities, should not Mercury's be more?

A nebular type theory, not necessarily following the pattern of La Place's, seems to have most appeal. In 1945, C. F. von Weizsäcker (1912–   ) proposed an original nebula with several inner turbulences developing into planets. The idea of Gerard Kuiper (1905–1973) was similar with separate clouds or proto-planets forming.

Since a conception about the development of the solar system is far from established, the question of why deal with the topic comes to fore. The best answer was given by the physicist Ludwig Boltzmann (1844–1906) discussing similar topics: "assuredly, no one will take such speculations for important discoveries, nor for the highest aim of science, as did the ancient philosophers. But it is not at all certain that it is right to laugh at them and treat them as altogether useless. Who knows that they do not enlarge the horizon of our circle of ideas, and even contribute to the advance of experimental knowledge, by increasing the nobility of our thoughts?"

SELECTED REFERENCES

R. Jastrow and A. G. W. Cameron, Eds., *Origin of the Solar System*, New York: Academic Press, 1963.

Thornton Page and Low Williams Page, Eds., *The Origin of the Solar System*, New York: Macmillan, 1966.

CHAPTER 9

ASTRONOMICAL INSTRUMENTS

A common, immediate association is telescope with astronomical instrument. Introduced by Galileo (1564–1642) at the beginning of the seventeenth century, the telescope is indeed a workhorse for the astronomer but other vital instruments are belittled by the common association. Among other important astronomical instruments are camera, spectroscope, photometer, blink microscope, balloon and artificial satellite.

The earliest astronomical instruments were probably globes. The oldest surviving example made of marble and with raised figures for the constellations, is in the Naples Museum, and dates from about 200 B.C. Armillary globes contained rings representing various circles of the celestial sphere such as the meridian and equator. Other models including those of the solar system called orreries came later. The first instrument used for measurement was the quadrant employed in obtaining star and planet positions. The device had a graduated arc and a pointer that pivoted about its center. With it, Tycho Brahe (1546–1601) was able to record positions of stars to within one minute or arc, equal to 1/30 of the moon's diameter.

The use of spectacles was the first application of optics in the western world. Several centuries later Galileo (1564–1642) introduced the optical telescope, with his first device, a three-power magnifier, being a couple of simple shapes of glass in a lead tube. The work of his telescope can be visualized today by holding a simple thick-in-the-middle lens at arm's length in one hand while moving a simple thin-in-the-middle lens in direct line in the other hand and moving the second lens until the best view is obtained. Later Galileo and others made more powerful telescopes.

Today's largest Galilean-type, or refracting, telescope is at Yerkes Observatory, Williams Bay, Wisconsin, not far from Lake Geneva. The front lens, or objective, is forty inches in diameter and consists of two parts, a front piece made of crown glass, 2½ inches thick in the center, and a rear piece eight inches away made of flint glass, two inches thick. The weight of the glass, twenty tons, indicates why larger refracting-type telescopes have not been built. Glass is a viscous fluid, not a solid, and can eventually sag; the lens is supported only at its circumference by the surrounding tube. Creative designers, aided by philanthropists, seeking to build larger refractors could employ clear and lightweight plastic, but the effort would be misguided because another equally-effective optical telescope can more easily be built in larger sizes.

The reflecting-type telescope was first suggested by James Gregory (1638–1675), a Scottish mathematician. At the age of 24, in his treatise *Optica Promota*, he described two concave specula, metal mirrors, with the larger one perforated. Construction was started but the telescope was never completed.

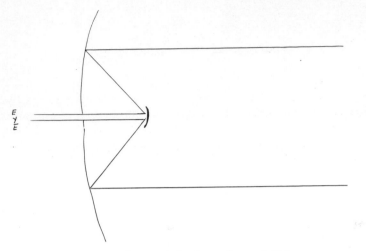

Schematic for James Gregory's Telescope

Several modifications were made in Gregory's basic plan after interest in a reflecting-type telescope was revived. That made by Isaac Newton (1642–1727) had a paraboloid objective and a flat mirror intercepting the reflected rays to send them to the observer. Shaping a portion of a sphere into a parabola is a tricky business but must be done if full value is to be obtained from the mirror. A spherical mirror does not return all parallel rays of light, especially those at the edge, to a common point, the focus, while a parabolic mirror does.

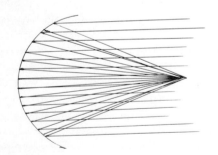
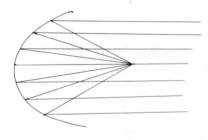

Sphere: Focus is Spread *Parabola: One-Point Focus*

Giovanni D. Cassegrain (1625–1712) in 1672 employed a convex secondary mirror placed inside the focal point of the prime objective. Such Cassegrainian designs are on many large, professional telescopes.

William Herschel (1738–1822) tilted the main mirror and the reflected rays went to the eyepiece at the other end of the tube. Herschel used a metallic mirror, as did the others, made generally of about 75 percent copper and 25 percent zinc. His largest reflector was about four feet in diameter.

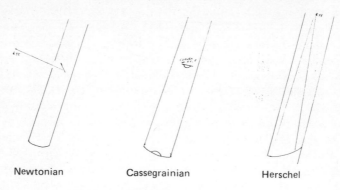

Newtonian   Cassegrainian   Herschel

Reflecting-Type Telescope Designs

The largest reflecting-type telescope prior to the twentieth century was built by an independently-wealthy Irish amateur astronomer, Lord Rosse (1800–1867). The device at Parsontown, Ireland, had a mirror six feet in diameter weighing about 8,330 pounds. It was very clumsily mounted because three men were required to move the telescope and keep the astronomer in position.

The mirror in Lord Rosse's telescope was made of metal. The idea of a glass mirror with a thin metallic deposit was discovered by chemist Justus von Liebig (1803–1873) in 1840 and first successfully applied in 1856. At the start, silver was used, later aluminum was favored and today a mixture of aluminum and beryllium called beral is deposited as a very thin coating.

The largest reflecting telescope of the first part of the twentieth century was, with a 100-inch mirror, at Mt. Wilson Observatory in Southern California. Later, a 200-inch reflecting telescope was installed at Mt. Palomar, not far away. These two, as well as the largest refractor at Yerkes Observatory, owe much to the organizing and fund-raising talents of astronomer George Ellery Hale (1868–1938). In 1968, construction started on two telescopes with 150-inch mirrors. One is in Arizona at the Kitt Peak National Observatory and the other is in north-central Chile at the Cerro Tololo Inter-American Observatory. During the early 1970s the USSR began to operate a telescope at Zelenchuksky in the northern Caucasus having a 6-meter, or 240-inch mirror, weighing 42 tons. Such an instrument could detect a tiny candle flame 24,940 kilometers away.

The likelihood of a larger reflecting-type telescope being built in the future is extremely remote. The expenditure of funds would be more judiciously applied to one of four other devices. A first possibility is a smaller telescope on a space platform, rocket or balloon. The National Aeronautics and Space Administration in 1966 planned four Orbiting Astronomical Observatories. The first failed almost immediately because of power supply malfunction. OAO-II, launched in a nearly circular earth orbit on December 7, 1968, lasted much longer than expected and retrieved much data with its four 12-inch telescopes and other devices. (According to U.S. plans made during the early 1970s, a 120-inch reflecting telescope will be in orbit about 400 miles up soon after 1980.) A second candidate for the money are image tubes or image multipliers. Similar to or actually television cameras or vidicon tubes these help collect more astronomical information per unit time. The third and fourth ways to use the money would be for Schmidt-type and radio telescopes.

The Schmidt telescope was first designed in 1930 by Bernhard Schmidt, a one-armed optician who was a "voluntary colleague" at the Hamburg Observatory. Instead of grinding a spherical mirror into a paraboloid, he made the change with a separate piece of glass in contact with the spherical mirror. He removed the plate, looking like a corrugated washing board surface, and shifted it several feet to the center of curvature of the sphere. The result was a telescope with a much wider field of view. The 48-inch Schmidt telescope-camera at Palomar Observatory can encompass about $6°$, represented by holding three fingers at arm's length against the sky.

Optical telescopes, refracting, reflecting and Schmidt, have one of three functions: magnification, resolution, and collection. Astronomers are not seriously interested in magnification unless they are students of solar system objects. Stars and other items remain relative pinpoints in all telescopes, no matter the magnification; the distance prevents any apparent enlargement. The large refractor at Yerkes Observatory has a potential magnifying power of 4,000 diameters but arrangements exceeding 1,000 diameters are rarely used. In all cases magnification follows the same formula as used for microscopes. The focal length of the objective, the distance from the lens or mirror to the converging point of parallel light, divided by the focal length of the eyepiece, the distance from it to its converging point for parallel light, yields the magnification. The resolving power of a telescope can also be formulated. For ordinary light the resolving power in seconds of arc is equal to 4.5 divided by the diameter of the objective in inches. The resolving power of the Yerkes Observatory 40-inch refractor is enough to see as separate entities two stars .1 second apart. The 200-inch Palomar telescope has a resolving power of 4.5/200 or .0225 seconds of arc. In other words, the Palomar telescope can see two objects that are a foot apart at a distance of 1,729 miles. The last, most important function of modern optical telescopes, is also related to size. The amount of light gathered is directly proportional to the size of the collector, lens or mirror. Professional telescopes constructed during the twentieth century were invariably for the purpose of gathering more light from distant objects and consequently were with increasingly larger diameters.

Practically all professional telescopes are used in conjunction with other instruments; the astronomer does not peer through the eyepiece. He uses a smaller auxiliary attached telescope called a finder to check on positions and objects. When eyepieces are on telescopes, they generally are made of two or more lenses; the front one is called the field lens and its function is to gather light and direct it to the eye lens. (The word comes from the Latin, meaning lentil because a double convex lens resembles the shape of a lentil.)

The five possible chief defects of lenses, including those used as objectives in refracting-type telescopes, are spherical aberration, chromatic aberration, distortion, astigmatism and coma. Spherical aberration is the failure of all parallel rays refracted by the lens to converge to one point. Chromatic aberration is the production of color after the parallel rays go through the lens. The doctrine by Isaac Newton (1642–1727) that glass of whatever kind had the same refractive index, accepted as gospel truth during his lifetime, prevented correction of the two defects. After his death, opticians began to combine lenses of different kinds of glass in order to correct the aberrations. Crown glass, used for windows, is made of silica, potassium or sodium oxide and lime. Flint glass is a denser glass with higher dispersive power and the silica in it was formerly obtained by pulverizing flint. The development of the German optical industry during the nineteenth century brought finer types of glass which aided in the removal of other lens defects. In distortion, the lens does not give a straight line image but instead produces it as a curved, negative or barrel distortion or as a positive, pincushion type. Astigmatism, really a product of a spherical aberration, is the occurrence when lines of the image in one direction are distinct while lines at right angles are indistinct. Coma, another type of spherical aberration, is a blur of light that partly surrounds an image formed by a lens.

Since its inception in 1609, the optical telescope has provided a wealth of information. In less than 400 years of use, the instrument has radically altered man's conception of the astronomical universe. At the start and throughout, the telescope had as revolutionary an impact as the heliocentric conception. In the early twentieth century a new kind of telescope began to be developed; it detected the other kind of radiation which the earth's atmosphere does not block, radio waves.

In 1931, radio engineer K. G. Jansky (1905–1950) studying background radio noise concluded that a sizable part of the atmospheric noise had an extraterrestrial origin. The first instrument exploiting this discovery was built in 1936 by Grote Reber (1911–   ) in his backyard in Wheaton, Illinois. In 1939 he produced the first radio map of the Milky Way at a wavelength of 1.87 meters.

Radio telescopes are more analoguous to photographic exposure meters than to optical telescopes and a better name for them would consequently be radio photometers. They do not give a replica picture of a sky area; they show the average radio radiation over the field of the aerial.

Radio telescopes penetrate much farther into space than do optical telescopes. The 200-inch instrument at Mt. Palomar picks up signals two billion light years away. The 210-foot radio telescope at Parkes, New South Wales, Australia can detect energy five billion light years away. A 1000-foot fixed reflector radio telescope has been in operation near Arecibo, Puerto Rico since 1964.

The 300-foot radio telescope at Green Bank, West Virginia.

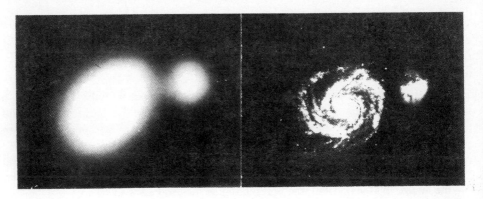

*A photograph of the galaxy M51 and the type of radio map of the same
that could be produced by a radio telescope.*

Radio instruments do not collect the entire band of radio energy but are restricted to particular wave lengths. Many of the radio sources at short wave-lengths coincide with photographic images while the long wave length items are not picked up by an optical telescope. The range of visible light waves is from 3500×10^{-8} cm. to about 7000×10^{-8} cm. On the other hand, radio waves span from 1,000,000 to about 1/10 of a centimeter. The Puerto Rico device covering 18.5 acres can operate between 6 and 1000 centimeters wave length.

Radio telescopes have a very poor resolution but the defect is being remedied by interferometer arrangements. One at the Fleurs Radioastronomy Field Station, 30 miles west of Sydney, Australia, consists of two intersecting arrays of telescopes, more than half a mile long apiece, each with 32 receivers, 19-feet in diameter, and two with a 45-foot diameter. The system makes possible a very detailed study of any source covering about one-ten-thousandth of a degree.

Radio telescopes need a great deal of auxiliary electronic equipment. Likewise optical professional telescopes are always used in conjunction with other devices. Among those commonly attached to the optical telescope are camera and photometer.

The camera can be traced to an early Greek entertainment device, a room light-tight save for a tiny hole. Sunlight enters this dark room through the small opening and the scene in front of the hole is reproduced upside down on the opposite wall. The image is upside down because light travels in a straight line from any point of the object through the small opening.

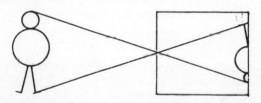

Representation of an Early Greek Entertainment Device

During the fourteenth century, the device was revived in Italy. Eventually, a small glass ball was placed in the tiny hole and this primitive lens sharpened and enlarged the picture image. The room was also reduced in size until the entire apparatus was as large as a shoe box.

During the nineteenth century, Louis Daguerre (1789–1851), a French theatrical scene painter used a sheet of copper and mercury vapor to make permanent, or fix, the image. In March, 1840, the first picture of the moon was so taken.

Today's cameras in astronomical observatories employ glass plates rather than the photographic paper known to the layman, but the coating is the same silver bromide imbedded in gelatin. Light striking silver bromide changes it to a different shade of color. Dim and distant objects are investigated by allowing their light to accumulate on the photographic plate over a great period of time. A permanent record of these time exposures is then made in the customary manner by developing, fixing and printing.

Before the advent of photography, brightness of astronomical objects was adjudged by the human eye, most sensitive to yellow-green light, roughly the region of 5500×10^{-8} centimeters wave length. The earliest photographic emulsions recorded the brightness of stars in the blue-violet region, about 4500×10^{-8} centimeters in wave length; the brighter the star the larger is the blackened area on the plate.

The most sensitive astronomical photometer is based on the photoelectric cell, where electric current produced is directly proportional to the amount of light entering the instrument. In the modern photoelectric photometer, a faint source of light can be amplified many times for accurate measurement. The drawback is the treatment of only one light source at a time whereas the photographic plate captures myriads of objects. Yet the photoelectric photometer is a vast improvement over the apparatus used in home, industry and picture-taking to detect how much light is available at various distances from the source.

Another important piece of auxiliary equipment for the astronomer in addition to the camera and photometer, but unlike the latter not attached to the telescope, is the blink microscope or comparator. It is employed to detect motion or change in brightness. The instrument enables a view through the same eyepiece of two seemingly-identical photographic plates of the same sky area. When the pictures are seen at rapid intervals, a flicker or blink is experienced should there be a change in brightness or position. Early motion pictures gave a comparable flicker because of poor editing or an abrupt change of scene. Also when toys were inexpensive, a popular one was a pack of paper or cards quickly flipped through the thumb to reveal a sequence of activity; flickers occurred at intervals.

Balloons and artificial satellites, used away from the observatory, can collect large amounts of data when outfitted with appropriate instruments. Balloon-borne telescopes into the atmosphere avoid the poor conditions of seeing through polluted, moving air.

In addition to a telescope and accessories, future satellites and space stations are certain to carry cameras and spectroscopes. These two instruments have not only made tremendous contributions to the development of astronomy but also they are used in practically every natural science from archaeology to zoology. Both devices began to be used in their present forms in the middle of the nineteenth century.

The central mechanism of a spectroscope is a wedge of glass called a prism or a piece with many thousands of ruled lines called a diffraction grating. When light from any source goes through either one, the energy is decomposed into its parts. In the case of the prism a rainbow of color appears while the diffraction grating produces many rainbows varying in brightness. With a school laboratory spectroscope the rainbows are visually observed; in commercial and research instruments, the rainbows are photographed.

Light from high-density material, solids, liquids and gases at high pressure, invariably yield a continuous rainbow after passage through the spectroscope. Light from a gas at low-pressure brings a series of bright-line spectra. The third kind of pattern, dark-line, also called absorption spectra, is of most interest to the astronomer. It is the product of light from a hot gas moving through an absorbing medium so that a vacancy, a dark line, registers where ordinarily a brightly-colored line is seen.

Since veritably all stars yield dark-line spectra, stars must have hot centers with cooler, surrounding atmospheres. This structure is as certain as other characteristics obtained via spectral analysis. Chemical composition, surface temperature, electrical condition, magnetic nature and relative motion, among other properties, are deciphered and known with the same degree of confidence.

Each of the little more than 100 chemical elements yields a unique array of light. The layman can visualize the fact by noting the yellow color of highway intersection lights, produced by sodium vapor burning, or the blue-green on some public roads due to mercury vapor. The human eye sees the dominant color in these and all other cases but the spectroscope registers the entire pattern, unique for each chemical element. Just as every man and woman on earth has a fingerprint like no other, so every chemical element has a spectral pattern like no other. Each brightly-colored line, in the red, orange, yellow, green, blue, indigo and violet region, is in a specific, quantitatively identifiable place. For this reason can an astronomer announce with assurance that an object millions of miles away has an abundance of hydrogen, small amounts of helium, and fractional percentages of boron and lithium.

Surface temperature is deciphered from the spectral pattern by noting which section of the plate has the most intense radiation. The common association of red with hot and white with hotter is a satisfactory guide, although the sequence from red to the other colors in the rainbow is a more sophisticated aid. When lines for a particular element are shifted to the red or violet end, the material in question is electrically charged. Should all the lines broaden, the source of light has a magnetic condition.

When two spectral plates of the same object are compared and all the lines shift to the red, or the violet end, the phenomenon determines increasing or decreasing distance of the light emitter. Called the Doppler effect after the discoverer Christian Doppler (1803-1853), the move towards the red or violet tells only whether the distance is becoming larger or smaller and not whether light emitter, spectroscope or both are moving. A shift of the lines, however small, to the red end indicates an increasing distance while the movement of the pattern to the violet means a decreasing distance. The phenomenon is the analog in light of a train whistle apparently increasing in pitch when approaching a station and having a decreasing frequency when moving from the station.

SELECTED REFERENCES

Gerhard R. Miczaika and Williams M. Sinton, *Tools of the Astronomer*, Cambridge, Mass.: Harvard University Press, 1961.

T. Page and L. W. Page, Eds., *Telescopes*, New York: Macmillan, 1966.

CHAPTER 10

CONSTELLATIONS

The Homeric epics and the writings of Hesiod refer to some star patterns but the earliest complete list of constellations is given in the poem *The Phenomena* by Aratus of Soli, published about 270 B.C. Forty-eight constellations are named. The group Coma Berenices was added to this list about 200 B.C. During the seventeenth century Johannes Bayer (1572–1625), Johannes Hevelius (1611–1687) and others gave names to sections of the sky seen only from the extremities of the southern hemisphere. About 100 years later Nicolas Lacaille (1713–1762) added thirteen more names to the southern hemisphere regions and subdivided Argo into four parts. Since then no new constellations have been added to the total of eighty-eight.

Modern people continue to call patches of the sky by such names as Canis Major and Orion despite the inability of many to visualize dogs, hunters and demigods formed by groups of the brighter stars. Johannes Bayer tried unsuccessfully to have the northern sky constellations named after New Testament figures and the southern constellations for people in the Old Testament. In 1950 a member of the British Parliament attempted to start a movement to change the constellation shapes and names but his efforts were short-lived and in vain.

Not many men and women, other than some amateur and professional astronomers, are able to identify all the constellations. The eighteenth-century British writer, Thomas Carlyle (1795–1881) wrote, "Why did not somebody teach me the constellations, and make me at home in the starry heavens which are always overhead and which I don't half-know to this day."

The ordinary observer looking up once a night must wait about a year to see all the constellations visible to him; another year at some other strategic point is necessary to see the remainder of the eighty-eight constellations. Seventy of the eighty-eight are visible, at least in part, in the United States; about a year in the southern hemisphere is required to view the remaining eighteen in the same cursory fashion. Observers willing to sacrifice sleep may see all the constellations visible at any one place by sighting three times in any one night: Early evening, near midnight and near dawn.

One of the best ways to learn to identify some of the constellations is to acquire a stock of patience in waiting for clear, night skies and to have a friend who knows the constellations point out some groupings. Many star guidebooks are also available.

Many of the constellations have associated myths, and the tales often indicate the hopes and frustrations of early peoples. There are several stories about the Pleiades, a part of a constellation seen with the naked eye as six or seven stars and described by the English poet Alfred Lord Tennyson (1809–1892) as "a swarm of fireflies tangled in a silver braid". The popular name of the Pleiades, the seven

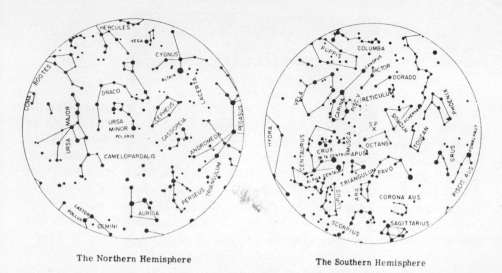

The Northern Hemisphere   The Southern Hemisphere

Some Major Constellations

sisters, stems from a Greek story involving the marriage of the giant Atlas and a nymph. The couple have seven beautiful daughters who were pursued by a warrior. The girls appealed to the god Jupiter who changed them to doves and they found refuge in the sky. The Iroquois Indians of North America told a tale about seven happy boys playing in the woods. The stars beckoned to the boys and they followed. One boy became homesick and started to cry, covering his face; thus six, and not seven stars are ordinarily visible to the unaided eye.

Another well-known story concerns six constellations in the northern skies. The cast includes Cepheus, a king, Cassiopeia, his queen, and their charming princess, Andromeda. The vanity of the queen enraged the gods who chained Andromeda to the sea rocks. Andromeda's boy friend, Perseus, a prince returned to the scene with the head of Medusa, the ugly one. Anybody viewing this ugly face directly turned to stone and Perseus had dealt with her through reflections from his shield. Perseus drew his sword to save his sweetheart and slew the menacing dragon, Cetus.

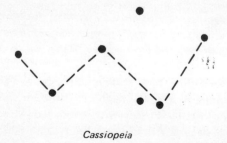

Cassiopeia

Courtesy: Lick Observatory

The galaxy in Andromeda.

The zodiac is the 18°-wide section of the sky containing the twelve constellations through which the sun apparently moves in one year. Each month the sun rises with a different constellation of the zodiac, and the sun's path, the ecliptic (where eclipses occur), is approximately in the center of the zodiacal strip. The twelve zodiacal constellations are described in the rhyme:

> The Ram, the Bull, the Heavenly Twins
> And next to the Crab the Lion stands
> The Virgin and the Scales;
> The Scorpion, Archer, and Sea-Goat.
> The Man that bears the watering-pot,
> The Fish with shining tails.

The zodiacal constellations usually called Aries, Taurus, Gemini, Cancer, Leo, Virgo, Libra, Scorpio, Sagittarius, Capricornus, Aquarius and Pisces, are the principals in the fake and erroneous correlations of astrology. This widely-followed entertainment is at variance with the established concept that heredity and environment govern an individual person's fortune and character and not the time of year in which born. The only good point for astrology is that it may induce some to look up and notice the stars.

Many intelligent men and women today cannot even recognize the north celestial pole star, Polaris. Seven stars called the Big Dipper, part of the constellation Ursa Major, form the guide to Polaris. The stars are about equally distant from each other and Polaris is five of these space distant from the closest of the two so-called pointer stars.

The Big Dipper

Great poets were always acquainted with the stars. The word is mentioned in all but four of the plays by William Shakespeare (1564–1616) who was very much taken in by astrology. His *King Lear* reports "It is the stars, the stars above us, govern our condition". In the same play is the comment "These late eclipses in the sun and moon portend no good to us". William Wordsworth (1770–1850) wrote:

> Look for the stars, you'll say that there are none;
> Look up a second time, and, one by one,
> You mark them twinkling out with silvery light,
> And wonder how they could elude the sight!

John Keats (1795–1821) is responsible for:

> Bright star, would I were steadfast as thou art —
> Not in lone splendour hung aloft the night.
> And watching, with eternal lids apart.

One of the contributions of Walt Whitman (1819–1892) is:

> Over all the sky — the sky! far, far out of reach,
> studded, breaking out the eternal stars.

Lesser-known Bert Leston Taylor (1866–1921) wrote:

> When quacks with political pills would dope us,
> When politics absorbs the livelong day,
> I like to think about the star Canopus,
> So far, so far away!
>
> To free, what I am pleased to call my mind,
> From matters that perplex it and embarrass,
> I take a glass, and seek until I find,
> A wisp of cloud, a nebula by name,
> Andromeda provides a starry frame.

SELECTED REFERENCES

Donald H. Menzel, *Field Guide to the Stars and Planets*, Boston: Houghton Mifflin, 1964.
Charles Whyte, *The Constellations and Their History*, London: Charles Griffin, 1928.

CHAPTER 11

STARS

It is commonplace to claim that the introduction of the telescope at the beginning of the seventeenth century ushered in a new era not only in astronomy but also in all modes of human thought. A similar claim can be made for every astronomical instrument. Alone, the telescope brought much data, but in association with the photometer or the camera or the spectroscope or the blink microscope, the information multiplied and new horizons appeared.

On a clear night in a pollutionless countryside the unaided human eye can make out about 3,000 stars. Today no one has a strong desire to check this figure, obtained by deciphering numbers in representative samples of the sky, because any size telescope reveals many more stars. The total number in the field of view of this, that or the other telescope can be ascertained. The telescope minus auxiliary equipment simply reveals stars in great abundance.

Together with an angle measuring instrument the telescope can yield information about the distance of some stars. The values are obtained by more than perfunctory tabulation; about 200 years elapsed after the first use of the telescope before a measurement was made.

Isaac Newton (1642–1727) speculated about star distances but his estimates were far too low. When he made his "guess-timates", 100 years of effort had already gone into the search for stellar parallaxes. As described in Chapter 3, advocates of the heliocentric system sought a small yearly, angular displacement of near stars against the background of the more distant ones. After the first parallax was measured early in the nineteenth century, obtaining star distances became more regularized. Parallaxes have been measured for thousands of stars but only those for about 700 are large enough so that the error is not too high.

The diameter of the earth's orbit, approximately 93,000,000 X 2 or 186,000,000 miles, is the base line in the procedure. The star in question is sighted from the earth at two- successive 6-month intervals in order to obtain two angles of a triangle. The third, the angle at the star, is had by subtracting the sum of the measured two from 180° since all three angles must total 180°. The diagram is misleading since the angle at the star is exceedingly small; when it is .01" or less, the procedure cannot be employed with any degree of confidence. The nearest star, beyond the sun, Proxima Centauri, has a parallax angle of only .78".

Any triangle where the value of three angles and one side are known can be ascertained completely through the elements of simple trigonometry. In this way the distance of the star to the earth's orbit is found to be 4.3 light years, 4.3 times the distance light travels in one year at the rate of 186,000 miles a second.

The distance is the same number whether it be to the earth, the sun, or Pluto. If, for example, 8-1/3 light minutes, the earth to sun distance be subtracted from 4-1/3

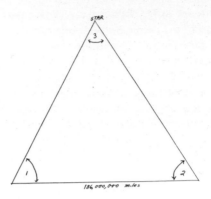

The First Step in Measuring Star Distance

light years, the answer is still in essence 4.3 light years.

If an object be considered present when the angle of parallax is 1″ then the distance corresponds to 3.26 light years. This value has been adopted as a new large unit of distance. Its description, parallax of one second, is contracted; the prefix of the two main words is joined to form parsec, a distance of 3.26 light years. The approximately 700 stars whose distance are found by the measurement of parallax are within 20 parsecs from the earth. The vast majority of star distances are found through other means and auxiliary instruments.

Relative brightness is another characteristic obtainable with a telescope and a single other device. Naked-eye observations gave the earliest scientists a few categories. The very brightest they called number one and the dimmest, barely detectable were number six. (Many societies still refer to their best merchandise as number one). During the middle of the nineteenth century, Norman R. Pogson (1829–1891) suggested the brightness, or magnitude, scale now in use after noting that a first magnitude star yields 100 times as much light as does a sixth magnitude one. The light sensation followed a geometrical progression and the magnitude designation an arithmetical one so that five units of change was responsible for the 100 units of light. The fifth root of 100, about 2.512, represents the amount of light difference between one magnitude and the next. A star of magnitude two gives 2.512 times as much light as does one of magnitude three, $(2.512)^2$ times as much as one of magnitude four, $(2.512)^3$ times as much as a magnitude-five star, and so on. A star of magnitude one gives the earth $(2.512)^5$ or 100 times as much light as does a star of magnitude six; a star of magnitude three gives $(2.512)^8$ as much light as does one of magnitude eleven.

Some magnitudes of well known objects are —26.5 for the sun, —12.5 for the full moon, —4 for Venus when brightest, —2 for Jupiter as well as Mars when brightest, and —1.4 for Sirius, the brightest star (aside from the sun) seen in the northern hemisphere. The other star with a negative magnitude is Canopus, seen in the southern hemisphere and as far north as the tip of Florida. When used visually the 200-inch telescope at Mt. Palomar can detect stars with magnitude twenty; the photographic limit of the device is magnitude twenty-four.

The apparent brightness or magnitude is scaled according to naked eye observation or photographic detection. The latter are measured on ordinary blue-sensitive photographic plates while the former is done with yellow-sensitive (orthochromatic) plates exposed through a yellow filter in order to roughly match the color sensitivity of the human eye. The retina is particularly sensitive to green, yellow, and red while the photographic emulsion is strongly affected by blue and violet. The difference between the visual and photographic magnitudes of a star is known as its color index.

Magnitudes may vary somewhat with observatories because of systematic discrepancies such as location of the observer. For example, the star Aldebaran has been listed as magnitude 1.5 as well as magnitude 1. It is now customary to correct magnitudes for the dimming effect of the atmosphere by reducing the data to that equivalent to outside the atmosphere.

The very brightest stars would be so whether viewed from a satellite or an observatory. There is complete agreement that the five brightest ones, in order, are Sirius or Alpha Canis Majoris with a magnitude of −1.58, Canopus or Alpha Argus with a magnitude of −.86, Alpha Centauri, +.06 magnitude, Vega or Alpha Lyrae, +.14, and Capella or Alpha Aurigae, +.21. (The stars may have a single name such as Sirius but one widely-accepted custom is to use the Greek alphabet to indicate brightness plus the genitive of the Latin constellation: Thus Alpha and Beta Delphinus, the first and second brightest in the constellation Delphinus, are also called Sualocin and Rotanev, being Nicolaus Venator spelled backwards. G. Piazzi (1746–1826) named the stars in this manner in order to recognize the services of his servant. Other names also creep into the literature. In the constellation of Cancer are Ascellus Borealis and Ascellus Australia, the Northern Little Ass and the Southern Little Ass. In Canes Venatici is Cor Caroli, meaning Heart of Charles, so named by Edmond Halley (1656–1742) for King Charles II of England.)

Some stars varying in brightness can be detected with the unaided eye. Betelgeuse (called Beetle juice through mispronunciation) in the constellation Orion is such a star. The first to be cited, in 1596, was Mira in the constellation Cetus; its maximum brightness is several hundred times its minimum brightness. Telescopic observation reveals many more stars varying in brightness. Delta Cephei, the fourth brightest in the constellation Cepheus, was so uncovered. However most of the approximately 20,000 stars recognized as variables were discovered through photographic and photoelectric techniques. A great number were found by using the blink microscope, comparing seemingly identical photographs of the same star taken on different dates.

To make a point, all stars can be considered variable since their light is effectively zero during daylight hours, or their radiation is truly reduced by the passage of a cloud in the earth's atmosphere across our line of sight. Since all stars twinkle apparently because of turbulence in the earth's atmosphere — they vary in brightness. Every star is also a variable because apparent brightness depends upon size, distance and temperature and these change with time. Nonetheless, astronomers reserve the term variable for the 20,000 cited above, any star whose light variations are the result of variable energy output.

The accepted method for naming variable stars is to place the letter R in front of the constellation for the first in the group to be discovered. After the letter Z has been used, two letters beginning with RR are employed. Should a constellation have many variables requiring more than the use of ZZ, then AA, AB through AZ are invoked, followed by BB to BZ, CC to CZ and so forth to QZ. This system allows for 334 variable stars per constellation; when more are available the designation is to use the letter V for variable plus numbers beyond 334 and the constellation name as V446, Cygni.

Several different classifications of variable stars are available. For one, they can be grouped according to period of pulsation into short, medium and long periods. In the 1958 edition of *General Catalogue of Variable Stars* issued in the U.S.S.R., most of the 15,000 stars listed are in the last category. Another popular classification is into extrinsic and intrinsic, with the former being variable because of conditions outside the star, such as another star regularly eclipsing its light. Algol is such a star. The Soviet handbook lists three types, pulsating, eruptive and eclipsing.

Pulsating variables having a short period of about half a day, often called RR Lyrae stars, after the prototype, are invisible to the naked eye; their maximum brightness places them only one magnitude away from their minimum brightness. On the other hand, long period pulsating variables, all red, large stars, have a wide range of magnitude. By far the most useful pulsating variables to astronomers are the Cepheid variables, yellow large stars named after the first to be detected, delta Cephei, discovered in 1784 by a young deaf-mute English astronomer John Goodricke (1764–1786).

For the Cepheid variables, the period of pulsation and their brightness at a standard distance, 32.6 light years, are directly proportional. This period-luminosity law was discovered by Henrietta Leavitt (1868–1921) at the Harvard College Observatory in 1912. Harlow Shapley (1885–1972) at the same institution was the first to see how the Cepheids could be used as distance indicators and pioneered in determining distances to some of them so that a period-luminosity graph could be established. With the latter, periods could be measured, luminosities read off the graph, and a calculation based on the fact that light falls off inversely as the square of the distance reveals the Cepheid star true distance.

Eruptive variables can be divided into dwarf novae, novae and supernovae on the basis of intensity of action which may not, in the last analysis, be a similar process. Novae are stars that increase in brightness from 5,000 to 100,000 times, dwarf novae increase in magnitude by a factor of 10 to 100, while supernovae acquire the Hollywood superlative appellation because in one second they burn with a glare equivalent to about 60 years of sunlight. Only three supernovae have occurred in our Milky Way stellar system during the last 900 years. One in 1054 was noted and recorded by Chinese astronomers; in 1572, Tycho Brahe (1546–1601) saw a star that at its peak was 5 to 10 times brighter than Venus; in 1604, Johannes Kepler (1571–1630) noted one that became as bright as Jupiter. However, supernovae are regularly found in other stellar systems and a study of the explosions has resulted in two competing theories. One claims the star has run out of "gas" and lost control of its heating mechanism and a surface blow-off is the consequence. The second idea holds that the entire star is disintegrating.

Ordinary novae flare up once, such as Nova DQ Herculis in 1934 and Nova CP Puppis in 1942. In rare cases, novae recur such as Nova T Coronae Borealis in 1866 and 1946, Nova RS Ophiuchi in 1848, 1933 and 1958, and Nova T Pyxidis in 1890, 1902, 1920, and 1944. After the nova, no matter the type, reaches its maximum brightness, an expanding shell of ejected material surrounds the star.

Expanding shells of gas have not been detected for dwarf novae. The intensity of their brightness is roughly proportional to the period between bursts. Stars reaching maximum brightness in ten to twenty days increase their luminosity by a factor of 15, those with a period between 50 to 70 days by a factor of 40, and those with a period of a year by a factor of 100.

Spectroscopic analyses reveal much about the eruptive variables — and about stars generally. The instrument employed in astronomy for little more than 100 years has put to rest philosopher August Comte's (1798–1857) notion that man would never know the stars. Without much sophisticated interpretation, stellar spectroscopy has revealed information about the chemical composition, surface temperature, electrical condition, magnetic condition and relative motion of stars, as outlined in Chapter 9.

Spectral analyses show that an ordinary star is from 50 to 80 percent hydrogen. Together with helium, the element makes up 96 to 99 percent of the average star. An unusual case, reported in 1948, is HD 124448, near the constellation Lupus, without a trace of hydrogen. Another highly remarkable spectrum is that of HD 101065 in Centaurus. Missing from it are the lines of iron, titanium, chromium, manganese and sodium that dominate the spectra of normal stars; instead the lines of the rare-earth elements are present. Another oddball is HR 4072, with an amount of platinum equal to the mass of the earth.

Whatever the chemical composition of a star, the material may be ionic, electrically charged atoms. The hydrogen present may not be the gas sold in steel cylinders but hydrogen ion, the atom minus a whirling, surrounding electron. Helium and all the others can likewise be in the same condition.

As outlined in Chapter 9, the spectral pattern of each element is unique and is recognized. When the element is ionic, the characteristic spectral lines are moved a bit from their customary position. The lines also reveal the surface temperature; wherever the radiation is most intense is the index to the numerical value. The range of surface temperature, disregarding exceptional stars, is small. The highest surface temperature is only about 10 times the lowest.

Chemical composition, surface temperature and the color of a star range in the

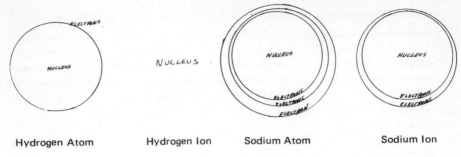

| Hydrogen Atom | Hydrogen Ion | Sodium Atom | Sodium Ion |

Representation of Atoms and Ions

same way. The spread of values of one is also true for the other two characteristics. The three fit the same way into the spectral classes first made by Jesuit astronomer Pietro Angel Secchi (1818–1878) in 1863 and now considerably expanded.

The chief spectral classes are O, B, A, F, G, K, M, R, N, S. The odd alphabetical sequence arises from the fact that A when discovered were thought to be the hottest and other stars would fall in line with the alphabet. However, the scramble of letters does form the first letter of every word in the enchanting phrase, "Oh be a fine girl, kiss me right now sweetheart". The sequence from O through S is from high temperature to low, from light elements through heavier ones and compounds, and from blue and blue white through the rainbow of colors to red stars. (The light from many stars enters the unaided human eye. The merge of colors gives the sensation of while although each star has a unique color.)

The Spectral Sequence

Class	Color	Surface Temperature	Chemical Composition	Example
O	Blue	25,000° K	Highly ionized & light atoms	10 Lacertae
B	Blue	11,000–25,000° K	Highly ionized & light atoms	Rigel
A	Blue	7,500–11,000° K	Hydrogen and neutral metals	Sirius
F	Blue-white	6,000–7,500° K	Hydrogen and neutral metals	Canopus
G	Yellow	5,000–6,000° K	Hydrocarbon radical present	Sun
K	Orange	3,500–5,000° K	Hydrocarbon radical present	Arcturus

(continued)

The Spectral Sequence (continued)

Class	Color	Surface Temperature	Chemical Composition	Example
M	Red	3,500° K	Titanium oxide present	Betelgeuse
R	Red	3,000° K	---	---
N	Red	3,000° K	---	---
S	Red	3,000° K	---	---

With a little ingenuity the spectroscope enables the determination of many other star characteristics. One that seems almost unbelievable to attain since all stars are no more than pinpoints in telescopes, is star size. The stellar interferometer and its electronic version have measured about two dozen star diameters, but the bulk of data about size of stars has come with the help of the spectroscope.

The first step in obtaining a stellar diameter is to measure the star's radiation by attaching a photometer to a telescope. The surface temperature is then found with the use of spectroscope and camera allied with the telescope. A well verified relationship, the Stefan-Boltzmann law yields the radiation per square centimeter from the surface temperature; E is the former in $E = kT^4$, where k is a listed constant and T is the surface temperature. The radiation of a star divided by its radiation per square centimeter gives its area, which is size.

The procedure results in values comparable to the few obtained with an interferometer. In both, Aldebaran (Alpha Tauri) comes out to be 45 times as large as the sun, Arcturus (Alpha Bootis), 23 times as large and Antares (Alpha Scorpii) 640 the size. Oddly, applying the procedure to the sun could give false values. The radiation per unit area could be found directly for our star and the unit area chosen might be a sunspot. Other stars no doubt have star spots but an average, rather than a particular radiation per square centimeter is found for them.

Likewise the mass of the sun is obtained somewhat differently than that for other stars. Gravitational attraction is the basis for the measurement in all cases but the sun's is with a planet and is more reliable while that for other stars is with a nearby star.

Two stars close, a binary system, is very common. In the neighborhood of the sun, from one half to two thirds of the stars belong to a binary or multiple star system; the sun is a real "loner". Observation of the two stars to determine the elements of their orbit and application of Kepler's third law relating time of revolution and average distance to the sun of the masses brings a value for mass. The amounts turn out to be in almost as narrow a range as surface temperature. Star masses vary from 1/10 to 10 times the sun, a 100-fold range, compared to surface temperature's ten-fold range.

One of the most varied star characteristics with huge, extreme values is a derivative of mass. By definition, density is equal to mass divided by volume. The average density of our sun, 1.4 grams per cubic centimeter, hides the fact that the density is huge at the center, as much as 80 grams per cubic centimeter, and very tenuous at the outer edges. Stars highly vacuous throughout do exist. Stars with highest average density are small, hot ones called white dwarfs, where one cubic centimeter is a matter of tons. For the companion to Sirius one average cubic centimeter weighs 1 ton; for Van Maanen's star, one average cubic centimeter weighs 7 tons; for AC +70°8247, one average cubic inch weighs 620 tons. Stars with even higher density are called neutron stars and are probably the pulsars, objects detected with the radio telescope during the late 1960s and sending regular periodic radiation. Some theorists point to even denser stars, not yet detected, called black holes.

The fact of extremely high density, not experienced on earth, calls for an explanation. On our planet, iron having 500 pounds to the cubic foot is considered dense and osmium with 1,500 pounds to the cubic foot is one of the densest elements; a white dwarf with a density of 10,000,000 pounds to the cubic foot is not unusual. The accepted explanation is that atoms on white dwarf stars are collapsed with the encircling electrons no longer at great distance from the nucleus but within the very center of the atom.

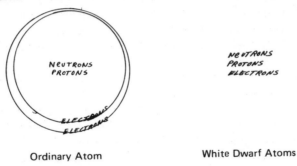

Ordinary Atom   White Dwarf Atoms

Planetary and White Dwarf Atoms

Perhaps black holes have all their atoms collapsed but white dwarf stars evidently have a supply of normal atoms on their surface. The movement of the electrons from one energy level to another accounts for the release of radiation. With electrons stuck in the atomic nucleus, there is no way for light and other energy to be emitted and detected on earth.

The first white dwarf to be found was suspected as a star following Sirius by Fredrick Bessel (1784–1846). Almost twenty years elapsed before the star was actually seen, by the son of Alvan Clark (1832–1897) the nineteenth-century American telescope builder. Bessel had been observing the motion of Sirius across our line of sight and credited the wave-like nature to an unseen companion of the star.

The tiny motion of stars perpendicular to our line of view is called proper motion and several other interfering movements must be substracted to find its amount. Diurnal motion, also across our line of sight can be ignored, but the effects of aberration, nutation and earth precession must be considered. Although several fairly large proper motions were uncovered before the introduction of the blink microscope, the comparison of seemingly-identical photos is now necessary.

The star with the largest proper motion, 10.3" per year, is Barnard's star, found in 1916 by the Nashville photographer E. E. Barnard (1857–1923) reputed to have extraordinary eyesight. Barnard's star is a 10th magnitude one in Ophichus. Another star with a large proper motion was discovered by J. C. Kapteyn (1851–1922) and has a movement of 8.7" per year.

The omnipotent, omniscient and omnipresent spectroscope enables measurement of a star's motion towards or away from us called the radial velocity. Most measured stellar radial velocities, via the Doppler principle, are between 10 and 40 miles per second but an unusual pair numbered 5583 and 5584 in the Washington Catalogue have a radial velocity of approach of 180 miles per second. Proper motion in one direction and radial velocity at right angles constitute the two most directly obtainable star motions.

Stars have many more movements but each of them is indirectly observed. Rotation, for example, is a fact for our sun not so much because the sun spots can be seen moving from one edge to another but because red shift and blue shift spectral patterns are simultaneously obtained from opposite edges of the sun. To conclude that all stars rotate because the sun does so is certainly not a safe induction. However, the spectral patterns of thousands of other stars have also been analyzed; if a star is rotating with sufficient speed, the lines are widened. Many stars are found to have rotation period of only a few hours. Deciphering of binary systems shows the revolution pattern of stars and in star clusters other kinds of movements can be delineated.

The phrase "fixed stars" is a misnomer even if the illusion of star motion is considered. Because the earth rotates, the stars and all heavenly bodies have diurnal motion, a turn once every 24 hours. Because the earth revolves, the near stars show the small shift against the background of more distant stars; such stellar parallax is not a real star motion and neither is the fact that the same star is seen in the same exact direction at the same time only once a year. Because the earth undergoes precession, our pole star apparently moves and again the motion is due to the earth, not the stars.

The real motion of stars helps attain other characteristics. In binary systems, star mass can be worked out; with proper motion, some star distances can be ascertained. A correlation between distance and proper motion reveals that the more distant the star the less its proper motion.

Astronomers and other scientists continually seek correlations. These are usually expressed in words or equations and occasionally in pictorial form as a graph. The relationship between proper motion and distance is a simple one, no matter the manner of expression; likewise the pattern is similar for mass and luminosity, with a one to one relationship. The correlation between luminosity and color, however, is not simple and direct and moreover its portrayal is in graph form only.

Luminosity, the magnitude at a standard distance of 32.6 light years or 10 parsecs, also has two other names. Sometimes it is referred to as absolute magnitude and occasionally as intrinsic brightness. Luminosity is, of course, the preferred, less-ambiguous name. Since color ranges in the same manner as spectral class, surface temperature and chemical composition, these may be used in place of color in any patterning employing the entire set of data. Whatever the terms used, pitting luminosity against color yields the Russell-Hertzsprung diagram.

The R-H or H-R diagram — the order of the alphabet depending upon such unscientific factors as prejudice — was independently found by Henry Norris Russell (1877–1957) and Ejnar Hertzsprung (1873–1967). They plotted the luminosity against the color of some stars and made a strange-looking diagram when the points representing the stars were connected. Instead of a regular figure such as a straight line circle or a parabola, easily translatable into words and equations, the R-H diagram could not be. When other astronomers added many more stars the same basic diagram remained.

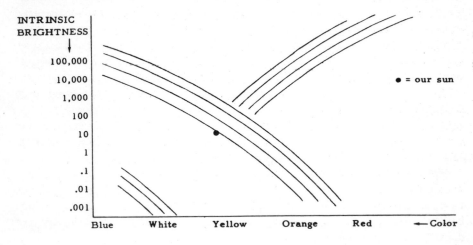

The Russell-Hertzsprung Diagram

Parts of the diagram were given unique names. Those stars encompassed by the longest portion were called main sequence stars; our sun is a main sequence star. Red stars in the upper reaches of the diagram received the title of red giants; white stars in the lower corners are white dwarf stars.

After serving as an interesting relationship for astronomy students, the H-R diagram took on new life with speculations about stellar evolution. According to presently held concepts, a star begins when gas, mostly hydrogen, heats up as it is compressed by gravitational forces. About ten million years may be required for the dust cloud to condense so that the pressure raises the temperature inside enough for the nuclear synthesis hydrogen into helium to occur. In 1936, a new star, since named F U Orionis, made an appearance in a concentration of dust and gas. It has been estimated that such an occurrence should take place within our view about once every 500 to 1,000 years. New stars have also been pinpointed as T Tauri variables, found in regions rich in dust and gas; more than 1,000 were found by the beginning of 1970. The T Tauri stars have thick and highly active outer atmospheres and are rapidly ejecting material into space.

The energy from the nuclear reaction finally comes to balance the force of gravitational contraction and the star becomes a part of the main sequence and may be for as much as 8 billion years. At first the star is not particularly bright; gradually it becomes hotter, bluer and brighter.

When the star has burned about 10% of its original hydrogen, it rapidly becomes brighter and redder. Within 100 million years at most, the star expands into the red-giant stage, destroying what planets it may have.

A red giant may pulsate for thousands of years before exploding into a nova and collapsing into a white dwarf. During the early 1960s an American astronomer estimated that half the stars in our galaxy are already white dwarf stars. As indicated earlier in this chapter the final fate of the white dwarfs could be neutron stars and at last, black holes.

Too long an immersion in the speculative about stars can be assuaged by practical stellar problems. One of these throughout time has been the cataloging of stars.

The first star catalogue of the western world is credited to Hipparchus (166 B.C.–125 B.C.) but not a single copy is known to exist. Should anyone ever turn up with the compilation by the father of Greek astronomy, rare book dealers would gladly pay a six figure amount for the volume. Claudius Ptolemy (90–168) probably copied much of the work in his own listing of 1030 stars.

Mohammedans and Asians kept alive the spark of science during the European middle ages but did not compile new star catalogues. From the Near East, Timur Lang, conquered most of the known world; his grandson, the Tartar ruler Ulugh Beigh (1394–1449) founded a large observatory in Samarkand, now the capital of the Uzbek Soviet Republic.

At the birth of modern science, star catalogs were prepared by Tycho Brahe (1546–1601), Johannes Bayer (1572–1625) who introduced our present method of naming a star with a Greek prefix and a Latin constellation, John Flamsteed (1646–1719), England's first astronomer royal, and Johannes Hevelius (1611–1687) who in 1690 compiled the first catalog of telescopic stars.

Later catalogs still referred to were developed by Charles Messier (1730–1817) with his 103 stars, clusters, and other objects; F. G. W. Struve (1793–1864) with his 3,110 pairs of double stars; and Henry Draper (1837–1882), a professor of chemistry with inherited money and the help of Annie Jump Cannon (1863–1941), a spectroscope catalog with about a quarter of a million stars.

One of the largest star catalogs was started by Friedrich Argelander (1799–1875) at the University of Bonn and so called the *Bonn Durchmusterung*. After all its extensions the catalog had more than half a million entries.

SELECTED REFERENCES

R. Jastrow, *Red Giants and White Dwarfs*, New York: Harper and Row, 1971.

R. J. Tayler, *The Stars; Their Structure and Evolution*, New York: Springer Verlag, 1970.

CHAPTER 12

GALAXIES

Many seemingly whiffs of gas appear in the skies to the unaided eye. Not clouds in the earth's atmosphere, these configurations can sometimes be magnified with telescopic view. At first all were called nebula, from the Latin for cloud, and differentiations were not attempted. During the early twentieth century, a subset called galaxies was firmly established but the name of nebulae for them still lingered. The result in our time is confusion in some circles because nebulae and galaxies look alike, even through the largest optical telescopes. Nebulae are unorganized gases between the stars while galaxies, despite the appearance of similarity to nebulae, are distant systems of billions of stars.

The first telescopic discovery of a nebula, in Galileo's time, was the giant one in the constellation Orion. Within a space of two years, the one in Andromeda, now known to be a galaxy, was cited. During the eighteenth century the French astronomer Charles Messier (1730–1817) searching the sky for comets, located 103 diffusely luminous objects and called them nebulae. However, 34 turned out to be galaxies and 57 were star clusters.

Nebulae can be seen as doughnuts of gas around a hot star, in a wide variety of shape, size and brightness, and finally as dark regions. The first are called planetary nebulae and more than 500 are known; the name arose because of their greenish appearance similar to Uranus and Neptune. A striking planetary nebula, inevitably displayed as an example, is the Ring nebula in Lyra. A pioneer in stellar spectroscopy, William Huggins (1824–1910), examining nebula, announced that the lines with wave lengths of 5007, 4959, and 3727×10^{-8} centimeters, represented a new chemical element and he named it nebulium. Not until 1927 was this mistake dispelled by the American astronomer Ira S. Bowen (1898–1973) who showed that unusual ions of oxygen and nitrogen produced the lines. Those nebulae, such as the large one in the constellation Orion, with a great variety of form, size and brightness, are called diffuse nebulae, and also exhibit brightline spectral pattern. Because of their spectral pattern both planetary and diffuse nebulae are called gaseous. In 1912, V. M. Slipher (1875–1969) at Lowell Observatory found that the diffuse nebula in Pleiades showed dark rather than bright line spectra; the nebula was reflecting the light of the embedded stars. The third type of nebula does not have enough light, either its own or reflected, to yield to spectral analyses and is appropriately named dark nebulae. When William Herschel (1738–1822) first saw them he said: "Here is truly a hole in the heavens". He and his sister Caroline (1750–1848) found about 2,500 nebulae and star clusters. His son John Herschel (1792–1871) extended the observations to the southern hemisphere and published a general catalog of nebulae and clusters; only 450 of the 5,000 entries were found by others than the Herschels.

The elder Herschel believed that many of the nebulae were what he called island universes. The same idea was also promoted by the English Instrument maker Thomas Wright (1711–1786), the German philosopher Immanuel Kant (1724–1804), the Swedish philosopher scientist Emanuel Swedenborg (1688–1772) and others lesser known. During the second half of the nineteenth century, with the introduction and use of the spectroscope in astronomy, nebulae were divided into two classes, green and white. The first were those such as the Orion nebula with bright lines and clearly gaseous; the second, the white, such as the one in Andromeda, gave dark-like spectral patterns and were therefore star-like. The appearance of a nova in the Andromeda nebula in 1885 seemed to be more evidence of its stellar nature. Also early in the twentieth century, the Andromeda nebula, M31, and the Triangulum nebula, M33, were resolved into some individual stars.

The characterization of those nebulae now called galaxies is chiefly the work of Edwin P. Hubble (1889–1953), an American originally trained to be a lawyer. Using the 100-inch reflector at Mt. Wilson Observatory he collected the information, codified it and finally theorized. Yet the title of his book *The Realm of the Nebulae* in 1936 added to the confusion of the nonspecialist about galaxies and nebulae. The latter are unorganized gases while galaxies, also occasionally called island universes, or spiral nebulae because of the shape so many of them have, or extra-galactic nebulae to indicate location, are systems of billions of stars.

The evidence that galaxies contain billions of stars is essentially the amount of light received from their distance. Either an entirely new phenomenon or billions of stars must be assigned. The dark-line stellar-type spectra of the galaxies and the resolution of a few into individual stars shows the star-like nature but gives no information about the number in any single galaxy.

Hubble used the period-luminosity relationship, outlined in Chapter 11, to find the distance of galaxies, since Cepheid variables were among the first stars delineated. His calculations showed the galaxy in Andromeda, the only one in the northern skies visible to the unaided eye, to be 800,000 light years distant. By 1951, Walter Baade (1893–1960) found two types of Cepheid variables with two different period-luminosity graphs. The appropriate one for distant galaxies gave the Andromeda galaxy a distance of 2,000,000 light years. The earth's distance to other galaxies reveals an average of a couple of million light years between one galaxy and its neighbor.

The best way to express the total number of galaxies is to say an uncountable large figure is involved. One estimate is that a trillion are within the range of the Palomar telescope.

Most of the galaxies entered in catalogs, about 80%, are spiral shaped, accounting for the phrase spiral nebulae as a replacement for galaxy. About 17% of the galaxies listed are elliptically shaped, small and faint, in contrast to the larger and brighter spirals. About 3% of the total documented have an irregular shape. Since Hubble's earliest classification of galaxies in 1925 other types have been found.

Early in 1940 Carl K. Seyfert (1911–1960) at Harvard College Observatory found about 10 spiral-shaped galaxies having very small, intensely bright centers and emitting an enormous amount of infrared and ultraviolet radiation. Now called Seyfert galaxies, they make up about 2% of the spirals.

Most galaxies congregate in groups. Our own is a member of the community called the Local Group and has at least 17 members. It includes the Andromeda galaxy and the two Magellanic clouds seen in the southern hemisphere of the earth. The latter are really suburban galaxies recorded in January, 1521 as the ships of Ferdinand Magellan sailed through the South Pacific by the chief chronicler of the voyage: "The antarctic pole has not the same stars as the arctic pole; but one sees there two clusters of small nebulous stars, which look like small clouds (nubeculae), at a short distance from each other".

The spectral patterns of all galaxies, except five close ones show a red shift pattern. This single piece of evidence prompted the promotion of an idea for the development of the universe. The so-called big bang or evolution theory proclaims that an original primeval explosion sent the galaxies hurtling into space and therefore registering red shift spectral patterns. Lately remnants of the radiation from the original big bang may have been detected. In 1965, A. Penzias (1933–   ) and R. Wilson (1936–   ) of the Bell Telephone Laboratories discovered a strange background microwave radiation coming from space; it is isotropic (arriving with uniformity from every direction) and has a black body temperature of only 2.7 degrees absolute.

The only competition to the big-bang theory, the steady state conception was forsaken by its chief promulgator. The idea would have the observable astronomical region develop only through what was called the creation of new galaxies. Continuous creation was a necessity of their cosmological principle: the universe looks very much the same from any location in all directions and at all times. To maintain a constant density in an ever expanding universe, matter must be created in space. The founder of the steady-state theory Fred Hoyle (1915–   ) when dropping it, called for an oscillation conception, an expanding universe now but contracting later.

In 1921 Edwin Hubble (1889–1953) established that the universe is expanding, with more distant galaxies receding faster. He calculated the ratio between the velocity and distance of a galaxy and it became known as the Hubble constant. In his time, the constant indicated an age of only 1.8 billion years for the universe, and consequently Hoyle and his co-workers were motivated to develop their steady-state model; the universe could not be younger than the earth, at least 4 and 1/3 billion years old. In time, the Hubble constant gave an age of 17.7 billion years and the expanding universe concept became more acceptable.

Astronomers are seeking verification of the Hubble constant at greater distances, using galaxies with large red shifts. Lately, huge red shifts of some quasi-stellar objects, quasars, have been detected and the values indicate an enormous distance for them. Found as a radio source at Ohio State University, OH 471 has a red shift of 3.4 (compared to a distant galaxy's .46) and the object, whatever it may be, is evidently about 11 billion light-years away. In 1973, astronomers at Lick Observatory in California reported the quasar OQ 172 in the constellation Bootes, even 50 million light-years more distant.

Quasars emit radio energy and have a star-like appearance. Their extreme red shifts places them at the edges of the observable region. They can help reveal whether the universe has continually expanded. Calculation of a constant called the deceleration parameter could indicate a universe that is decelerating and eventually contract.

If the universe is contracting the effect on our own galaxy should not at first be catastrophic. Probably some dimensions would change.

Information about our own galaxy has been in less state of flux than data about other island universes. However, at the beginning of the twentieth century the diameter of our own galaxy was thought to be 7,000 light years; by 1915 it was described as 15,000 light years; now it appears to be about 100,000 light years wide at its greatest thickness. The dimensions are found by adding the distances of the farthest stars 180° apart. Those along the Milky Way, the edge of our galaxy, represent the longest dimension while the values at right angles sum up to the thickness. Larger telescopes may yet find stars more distant belonging to our galaxy.

The sun is near the principal plane of the system but is 30,000 light years from the center in the direction of the constellation Sagittarius. The North Pole of our galaxy is in the direction of the constellation Coma Berenices and the opposite pole goes through the direction of the constellation Sculptor.

Our galaxy, spiral-shaped, rotates once every 200 million years. In the sun's vicinity the rate comes out to be about 600,000 miles per hour, or more than 160 miles a second, as indicated in Chapter 3. The Dutch astronomer J. C. Kapteyn (1851–1922) was a pioneer in estimating the shape and size of our galaxy. At the beginning of this century he measured the brightness and position of almost 500,000 stars. He worked alone although he once had the help of some convicts on loan from a prison.

The number of stars in our galaxy, about 50 billion, is estimated by counting the stars in sample patches of the sky and multiplying by the number of such sections. Since the result will vary with the size of instrument and investigator, published figures vary from 10 to 100 billion stars, with vast distances between them. Should an average star in our galaxy be reduced to the size of a raindrop, the average distance between stars would be reduced to about 40 miles.

SELECTED REFERENCES

D. W. Sciama, *Modern Cosmology*, New York: Cambridge University Press, 1971.
G. J. Whithrow, *The Structure and Evolution of the Universe*, New York: Harper Torchbook, 1959, Rev. Ed.

CHAPTER 13

TIME

One of the earliest applications of astronomy was in the measurement of time. The first societies, east or west, were not as slavish to the passage of hours as we are but were nonetheless interested in noting the duration of daylight, the period from new moon to new moon, or the time from one lunar eclipse to the next. The events in the sky seemed regular and reliable, although some tribes of men must have been equally impressed with the periodic flight of birds, the hibernation of animals, or even human cyclic phenomena. Astronomical events were not a necessary basis for measuring time but theological and practical factors established the tradition. Clepsydras, contrivances marking off intervals through the steady flow of a known amount of water, were used in China as early as 4,000 B.C., whereas in the western world measurement of duration was by means of heavenly phenomena. Perhaps the latter gave an aura of majestic truth, desired by the powers in Church and State. Both establishments proclaimed and modified calendars presumably synchronized with one or more astronomical events.

If a choice had to be made today of a regularly periodic heavenly phenomenon, the rotation of the earth would probably be selected, as in the past, over lunar eclipses, phases of the moon, or variability in star brightness. The rotation of the earth is not as regularly periodic as the vibration of certain molecules, now a time standard at the United States National Bureau of Standards, but regular enough for practical use. If the rotation of the earth were to be deciphered today, the best procedure would be to use the Foucault pendulum, described in Chapter 3. The ancients had to obtain the measure of the earth's spin through the sun and stars.

The interval between two successive passages of a star across the meridian is called a sidereal day. One complete earth rotation has occurred during this period and is reflected by the diurnal motion. Astronomers are content to live with sidereal time but other men and women want to guide their activities with our main star, the sun.

The interval between two successive passages of the sun across the meridian is called an apparent or true solar day. One complete earth rotation has occurred during this period and is reflected by the sun's diurnal motion. The measurement at the center of the sky, the meridian, is easiest and best. A dawn to dawn, or dusk to dusk interval raises the issue of when citings should occur — the first or last glimpse or even middle of the sun; moreover, horizons vary considerably so that the investigator at a coastal location has a different value than one in an urban area.

The earth rotates regularly enough to expect apparent solar days to be equal — but they are not. The difference between any two successive days is not much; that between the longest and shortest apparent solar days of the years is 61 seconds, a fraction more than a minute of time. The differences can be assigned to earth revolution and plane of travel. The earth's revolving about the sun at an irregular speed interferes with the measurement of rotation: A given spot on the earth is sunned more often when the earth is closest to the sun, since the earth is traveling faster about the sun, even though earth rotation time is equal throughout the year. The orbit plane also interferes because the earth travels above and below the central plane of the sun.

The interval between one apparent noon and the next is too irregular for practical use and an ingenuous solution was devised. The duration of 365 apparent solar days was averaged for a mean solar day. A "mean" sun was invented that crossed a given meridian at uniform intervals throughout the year. To make the system completely practical, correction factors to be added or subtracted to the apparent solar day to yield the mean solar day were compiled. Called the equation of time, the list gives the necessary amount, up to sixteen minutes, for an observer to use. Astronomers accustomed to sidereal time can also convert to mean solar time because the sidereal day is four minutes shorter than the mean solar day; 366¼ sidereal days are equal to 365¼ mean solar days.

The time we live by, civil time, is mean solar time. Two P.M. indicates that the mean sun is two hours past meridian while eight A.M. signifies eight hours before or ante meridian. All points on the surface of the earth could have the same mean time at a given instant. Astronomical observations do not prevent Chicago, New York, Sao Paulo, Moscow, Peking and Saigon having two P.M. simultaneously. Any introduction of such a system would bring an immediate uproar because men, women and children everywhere want to synchronize their activities with the understanding that the sun on the meridian is noon. Time zones were invented for this purpose.

The earth is divided into 24 time zones, each fifteen degrees wide. The value results from the rotation speed of 360° in 24 hours or 15° an hour; every 15°-span of the earth can identify its noon with the sun on the meridian. Within continental United States there are four time areas and a small fraction of a fifth. Eastport, Maine has Atlantic time while the remainder of the 48 states use the eastern, central, mountain and Pacific values. The wider expanse of the U.S.S.R. has nine time zones. Investigators using research stations close to the poles of the earth have the option of 24 different time zones but generally employ that of their home country.

Before travel became popular and extensive, the average person was not concerned with the hour at different places on the earth. Americans and Europeans could point to darkness and daylight being reversed in Asia, or Japan could be referred to as the land of the rising sun. The steamship, railroad, automobile, and airplane made people more aware of varying time. Today even small children know that New York City time is one hour more than Chicago time; that Denver has one hour less than Chicago and San Francisco is two hours behind Chicago.

In theory time zones everywhere are 15° wide but in practice the width varies because of local and regional ordinances. Separation lines between one area and the next are jagged because residents legislate themselves into a time zone irrespective of the natural boundaries. The internationally-agreed-upon first area beginning at London, England, has been so affected and likewise the ones twelve time zones distant marking the International Date Line.

By accepted convention and law, dates everywhere change at midnight but the time zone affair also necessitates a change of date at the International Date Line. Crossing the Date Line when moving east brings a smaller number while traveling westward yields a larger date. The process is just opposite to that in going from one time zone to another. United States sailors sensitive to pay gain and losses coined the phrase that best describes the events: East is least and west is best.

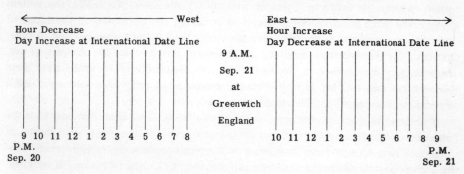

Time Zones and Date Line

The date changes at the International Date Line regardless the calendar used. The latter are also a matter of convention, in this case adapted to approximate an astronomical duration such as the time required for the earth to revolve about the sun. The Mayas used the time for Venus to go about the sun, about 584 of our days while the American Indians had the moon's period about the earth as a base.

The ancient Egyptians were among the first to introduce the solar-year calendar, beginning their year when Sirius rose with the sun. The time was divided into 12 equal months of 30 days, and five days of holiday were between one year and the next. Many societies with calendars based on the moon were gradually won over to a solar calendar although religious groups in all countries still maintain the former. The Mohammedan year has 12 lunar months, each 29½ days long, with a single 30-day month every 11 years and a single 29-day month every 19 years. The Jewish calendar has five lunar months of 30 days, five months of 29 days and two months varying between 29 and 30 days. An extra month of 30 days is intercalated every 3, 6, 8, 11, 14, 17 and 19 years in order to synchronize with the solar year.

Julius Caesar (100 B.C.–44 B.C.) was instrumental in the adoption of the solar calendar. Impressed with the ability of the Egyptian priesthood to predict the time of the Nile floods he "escorted" astronomer Sosigenes to Rome and had him develop a calendar. The Julian calendar had 365 and ¼ days divided into 12 months, with July named after Caesar. With his assassination in 45 B.C., the Senators changed the name of the month Sextilis to August in honor of Augustus Caesar

(63 B.C.–14 A.D.) the new ruler. In the Julian calendar the uneven-numbered months considered lucky, had 31 days and others had 30 days. In the Augustan calendar, a day was taken from February and added to August in order not to offend the ruler; to avoid three months in a row with 31 days each, a day was taken from September and November and given to October and December.

The present civil calendar, the Gregorian, after Pope Gregory XIII (1502–1585), came to be in 1582 when 11 days was omitted from the Augustan calendar in order to have a better synchronization with the true astronomical year. Men and women rioted demanding their 11 days; also about 200 years later, in 1752, when Great Britain adopted the calendar. It was Japan's turn in 1873, China's in 1912, Turkey's in 1917, the U.S.S.R. in 1918 and Rumania in 1923.

Some users of the Gregorian calendar may not be aware of its nature and character. The true astronomical year is 26 seconds shorter than the Gregorian one. The latter is best described by the kindergarten ditty about 30 days hath September and so forth, or the Girl Scout rule about knuckles on a fist representing 31-day months and the depressions being months with the smaller number of days. During years divisible by four or century years divisible by 400, an extra day is added to February.

One of two calendars now devised may in time supplant the Gregorian. One is the Julian Day Number, attractive to astronomers and historians and the other is the World Calendar, promoted by many more although a contemporary American, Elizabeth Achelis (1880–1971) has been a driving force for its adoption.

The World Calendar has 364 days divided into 12 months and 52 weeks but its unique aspect is four equal quarters, each 91 days long and composed of a 31-day month and then two, 30-day months. Each quarter begins on Sunday, with holidays always occuring on the same day. Artists, models, calendar printers and some religious groups are less than enthusiastic about the World Calendar while accountants, record keepers and some internationalists favor the plan.

The Julian Day Number was devised by Joseph Scaliger (1540–1609) in 1582 and named in honor of his father. He found the common multiple of three well-known time cycles, the solar, Metonic and Roman indiction, to be 7,980 years. The first cycle is the 28-year recurrence of the same day of the week on the same day of the year; the second is the 19-year cycle approximately equal to 235 lunar months when the new moon recurs on the same days of the year; the last period has no astronomical significance and is the 15-year Roman Empire periodic valuation of property for assessment. The first Julian period begins January 1, 4713 B.C. and goes to 3267 A.D. Each day from the beginning is called in order by number with the date changing at noon. Thus, January 1, 1935 is simply 2,427,804. Chronologists dating the exact time of an event or astronomers intrigued with date change at noon rather than midnight would find the Julian Day Number appealing, but for the average person the plan would be equivalent to a social security number being a calendar date.

SELECTED REFERENCES

Harrison J. Cowan, *Time and Its Measurement*, New York: World, 1958.
Beulah Tannenbaum and Myra Stillman, *Understanding Time*, New York: McGraw-Hill, 1958.

CHAPTER 14

PLACE

A second classical application of astronomy, in addition to the delineation of time, is the finding of exact earth location. The position of the stars can help determine the place of an observer on the earth. The one basic principle involved is agreement on standard references and measurements.

Finding one's way in the city of Chicago, or any large urban area, illustrates the fundamental idea. In Chicago, Madison Street and State Street are the agreed references for location of any spot. A house that is 7200 north and 1600 west must be 72 blocks north of Madison Street and 16 blocks west of State Street. Another system of location, also used in astronomy, is not as accurate. When a building is said to be in West Rogers Park, Woodlawn or Austin, the place is cited in the same general way as when an astronomical object is said to be in the constellation Andromeda or near Orion.

A location on earth is pinpointed with references to the imaginary plane called the equator as well as the one labeled Prime Meridian. The distance from the equator can be anywhere from $0°$ to $90°$, north or south; the distance from the Prime Meridian, the line joining the north pole, south pole and London, England, can range between $0°$ and $180°$, east or west.

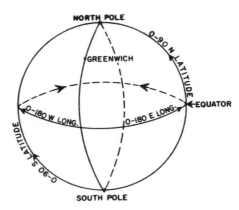

References on the Earth

The measure from the equator is called latitude and the one from the Prime Meridian is termed longitude. Latitude references are more frequent in public affairs with the seventeenth parallel separating North and South Viet Nam, the thirty-ninth north parallel dividing North and South Korea and the forty-ninth degree north of the equator being the boundary between much of Canada and the United

States. Longitudes are useful in deciphering time differences since 15° of longitude is equivalent to one hour of time. Between 75° west longitude and 150° west longitude there are 75° difference; 75° divided by 15° is 5, so 150° west longitude is 5 hours earlier in time than is 75° west longitude. Between 60° east longitude and 60° west longitude are 120°, eight hours difference in time, making places at 60° east longitude eight hours later than those 60° west longitude.

Latitude and longitude characterize earth positions uniquely. Thus Chicago has a latitude of 41°53'1" and Rome has a comparable north latitude of 41°53'34" but the longitude of Chicago is 87°36" west longitude and Rome has an east longitude of 12°29.7'. Stockholm is at 18°3.5' east longitude and Cape Town, Africa is at 18°28.7' east longitude but Stockholm has a latitude of 59°20.6' north while Cape Town is at 33°56.1' south latitude. The only place on the surface of the earth having zero latitude and zero longitude is on the Gulf of Guinea on the west coast of Africa.

A globe is the best way to represent the earth, showing true perspectives of latitude and longitude, but large globes have been more decorative than scientifically useful. One eleven feet in diameter and thirty-four feet around the equator is in the National Geographic Society's exhibition hall, Washington, D.C. A rotating, two-ton, aluminum globe, twelve feet in diameter is in the New York Daily News building. One of the largest, 60 feet in diameter, also holds natural gas at Savannah, Georgia. In 1824, a globe 128 feet in diameter was erected in the Champs Elysees, Paris. The so-called unisphere, 120 feet in diameter, was the permanent symbol of the 1964–65 New York World's Fair.

Maps try to show the surface of the earth but not a single known method of map making can indicate true distances, shapes and areas at all times. Latitudes and longitudes are distorted in every approximation. One general solution to project, or transfer point for point, the sphere representing the earth onto an enclosing cylinder, the Mercator projection, is only accurate where the sphere and cylinder contact. Should the contact be the equator, then Greenland appears to be larger than South America although the latter is 9½ times the size of Greenland. Alternative projections, such as a sphere onto a cone or the sphere onto a plane, bring other difficulties.

Those interested in determining latitude and longitude generally carry maps rather than globes. Should their position on earth be determined with the help of the stars, other devices and other knowledge must be available. Two systems of location on the sky must be known, the horizon and the celestial equator systems. In both, the sky is termed the celestial sphere and is considered part of the imaginary globe, at infinite distance, containing the entire universe. In both, the observer on earth is imagined, for calculation purposes, at the center of things.

The horizon system is so named because one reference is the observer's horizon, essentially the eye-level of the normal, upright person. The zenith point is 90° from the horizon, directly overhead the observer; the point 90° below the horizon is the nadir. No one seeks much below his horizon so that the nadir is a theoretical construct and a word valuable to poets. Strictly speaking, the horizon is the perpendicular bisector of the line joining the zenith and nadir.

In the horizon system, the altitude of an object is its arc distance above the horizon and may be any number from 0° to 90°; the altitude of the zenith is 90°. The altitude of any object locates it on a circle parallel to the horizon just as the latitude of a place on earth is essentially on a circle parallel to the equator. Just as the latitude of 41.5° north can be Chicago, Rome and any number of other stations, so a given altitude is indefinite.

In the horizon system, altitude must be coupled with azimuth to pinpoint an object to a particular spot. The azimuth, or bearing, measurement is equivalent to finding an exact direction; it is also comparable to longitude in the earth system. Astronomers measure azimuth from the south point while some surveyors and navigators start from the north point. In any event, the definition must be borne in mind. In general terms, azimuth is the arc distance, measured on the plane of the horizon, from the reference point to underneath the place in question.

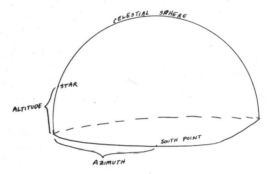

Essentials of the Horizon System

The south or north point is indicated by the earth's geographic poles, not the earth's magnetic poles. A magnetic compass points to magnetic north and south, both some distance from geographic north and south.

The cardinal points can be ascertained from astronomical observations. During the day, the sun on the meridian points to geographic north and south; when the meridian bisects the sun, the central line of the sky also meets the horizon at geographic north and south. During the night, a similar bisection of the diurnal arc of a star also yields the same cardinal points; or imagining a perpendicular from Polaris to the horizon delineates the north point.

The altitude and azimuth of an object can be estimated visually but a measurement with appropriate instruments is more meaningful and useful. A telescopic sight on a graduated circle is necessary for obtaining azimuth quantitatively. Altitude is measured with a sextant or octant, optical devices designed specifically for determining distance above the horizon.

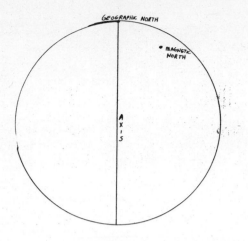

Magnetic and Geographic Poles

Azimuth is a valuable piece of data because it is essentially the direction. Altitude is also important. For several thousand years the altitude of the North Pole star, Polaris, has been a guide to navigators. Some early peoples knew that the altitude of the north celestial pole is equal to the latitude of the observer. Below is a simple proof.

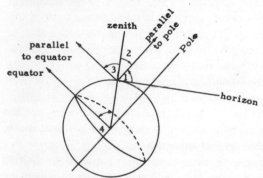

Angles $1 + 2$ are $90°$; likewise angles $2 + 3$ are $90°$. In the former case, the distance is from horizon to zenith and in the latter instance the angles are from equator to pole. Since $1 + 2 = 2 + 3$, angle 1 = angle 3. Angle 3 is equal to angle 4 because in Euclidean geometry corresponding angles formed by the intersection of two parallel lines and a third line are equal. Therefore, angle 1 is equal to angle 4, or the altitude of the north celestial pole is equal to the latitude of the observer.

Altitude of North Celestial Pole Equals the Latitude of the Observer

The north rather than the south celestial pole is involved in the dictum because man flourished in the northern hemisphere. (The southern hemisphere, relatively recently established as another center of civilization, can use the middle of the Southern Cross as a comparable guide.) The early Hawaiians had a device called the Sacred Calabash, now in the Bishop Museum, Honolulu, that determined when they reached the latitude of the island of Hawaii in the return voyage from Tahiti. The

Sacred Calabash was constructed so that Polaris was seen through it when the altitude of the pole star was 19°30', the latitude of Hawaii.

Simultaneously obtaining both components of position on earth, latitude and longitude, is expedited by a method of celestial sphere location called the celestial equator system. For astronomers, the latter is also more useful than is the horizon system.

The fundamental references in the celestial equator system can be visualized at first by again considering the stance of the earth in its orbit around the sun. Note that the earth's equator and the earth's orbit are at an angle of 23½°.

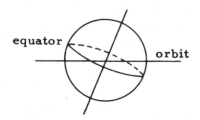

The Earth in Orbit

The equator and the orbit can be imagined stretched to the celestial sphere, whereupon they are called the celestial equator, or equinoctial, and the ecliptic. The latter is the term because eclipses occur within its bounds; the ecliptic is the midline of the zodiac, the 18°-wide strip through which the sun, moon and planets appear to travel. The celestial equator may be found in any sky by connecting a few points that are 90° from either the north celestial pole or south celestial pole. If the two circles are turned 23½° so that the celestial equator is perfectly horizontal, and the celestial sphere encloses both then the appearance is amenable to better understanding.

The two points where the celestial equator and ecliptic intersect are called equinoxes. One is the autumnal equinox and the other which occurs about six months later is the vernal equinox. Equinox days, in latitudes between the Arctic and Antarctic Circles, are those two days of the year, about September 21 and March 21, when the sun rises due east and sets due west. On equinox days, the duration of daylight equals the duration of darkness; the noon sun is exactly 90° above equatorial horizons; the same amount of sunshine is north and south of the

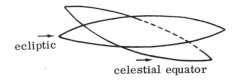

Celestial Equator and Ecliptic

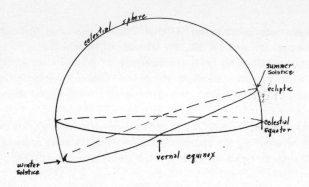

The Bases of the Celestial Equator System

equator. A couple of thousand years ago, the vernal equinox was close to the constellation Aries; the first point of Aries and the vernal equinox were synonymous — and still are in the fakery of astrology. The vernal equinox has shifted to Pisces because of earth precession, detectable at the poles as well as at the equinoxes. The continual change in earth-axis direction, described in Chapter 3, and the continual shifting of the intersection points called equinoxes are two manifestations of the same phenomenon.

Not only are the equinox points representative of dates but also the entire ecliptic represents a yearly calendar. About March 21 the sun rises due east and each day thereafter shifts a bit to the north. The sun rises farthest north of east about three months later, June 21, the summer solstice. From about June 21 to September 21, the sun's point of rise is still north of east but the amount decreases daily for three months. Between September 21 and December 21, the winter solstice, the sun has a point of rise that gradually shifts to the south. On or about December 21, the rise and set of the sun is farthest south of east and west. Between December 21 and March 21, the sun slowly regains the due east and west position for rise and set.

As in the earth and celestial horizon systems, two values specifically pinpoint an object in the celestial equator system. The arc distance above or below the celestial equator, ranging from zero to $90°$ is called declination while the arc distance from the vernal equinox, to the right along the celestial equator, to the place under or above the object is termed right ascension. (Those who prefer to measure leftward from the vernal equinox use the phrase, hour angle.)

If Polaris is considered at the north celestial pole rather than a short distance away, its declination and right ascension is $90°$ and indeterminate, regardless the day of the year. The sun on March 21 has a declination of $0°$ and a right ascension of $0°$; on June 21, the sun's declination is positive $23½°$ and right ascension amounts to ¼ of a circle or $90°$. On September 21, the sun's declination and right ascension is $0°$ and $180°$; on December 21, the declination of the sun is negative $23½°$ and the right ascension is three-quarters of a circle or $270°$.

Mercury is generally very close to the sun so that the values of declination and right ascension for our star could possibly hold true for the smallest planet. The largest error thus made in declination would be one-half the zodiacal path region, one-half of 18° or 9°, since the ecliptic is in the center; the maximum error in right ascension would be the maximum distance Mercury is from the sun as seen on our sky, or 28°. The declination and right ascension of Mercury on April 21 could be assigned therefore as 1/3 of 23½°, or 7.8° and 1/3 of 90°, or 30°. Likewise the declination and right ascension of Venus on May 21 could be close to 2/3 of 23½°, or 15.6°, and 2/3 of 90°, or 60°. In the case of Venus, the maximum error in declination is still 9° but the possible largest error in right ascension is 47°.

Celestial objects are located through their declinations and right ascensions as listed in star catalogs or navigational aid. For astronomers the information is simply the direction in the sky where the object is precisely located. For navigators, the data helps locate the individual's latitude and longitude on earth.

The problem of finding position on the surface of the earth can be resolved through many means. Piloting is a procedure using buoy and lighthouse sightings, soundings, radio directions and charts. Deduced reckoning, or "ded" reckoning attempts the problem with compass, log book and distance measurements through engine revolution readings. Loran, or long range navigation, permits guidance via radio. Celestial navigation is finding one's position on the earth by using clocks and astronomical observations. When first accomplished completely and successfully, during the nineteenth century, it was called an epoch discovery.

Several prerequisites are necessary for celestial navigation. The geographical directions must be ascertained, whether day or night. The navigator must be familiar with the approximate position of about 55 brighter stars as well as the planets observable with unaided eye. He has an accurate timepiece, a chronometer. He can use the sextant to find altitude and the radio sextant to track the sun or moon when the sky is obscured. He is familiar with essential books of data called almanacs.

The American Ephemeris and Nautical Almanac is issued by the United States Naval Observatory two or three years ahead of possible use. *The American Air Almanac* is a more complete listing of declination and right ascension, at ten-minute intervals for sun, moon and planets. Among the other useful volumes issued by the Hydrographic Office of the United States Navy is a revision of Nathaniel Bowditch's *American Practical Navigator*. Bowditch (1773–1838) was a sailor until the age of 30; he also sold insurance, became a trustee of Harvard University, and translated *Systems of the World* by Marquis de LaPlace (1749–1827).

Celestial navigation involves drawing two circles of position with two points of intersection; one of the latter yields latitude and longitude. Another procedure is to use the ZPS (zenith, pole, star) triangle outlined below.

The arc distance of the star to the horizon is its altitude and zenith to star is 90° minus the altitude. The arc distance of the star to the celestial equator is its declination and pole to star is 90° minus declination. The altitude of the pole is equal to

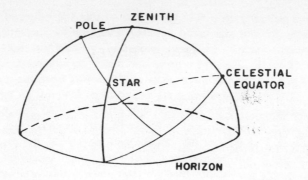

A Simple Navigation Problem

the latitude of the observer so that the arc distance ZP is 90° minus the latitude. The three sides of ZPS triangle are therefore established and the longitude is obtained through mathematical formulae involving the angle at the pole.

SELECTED REFERENCES

John Favill, *Primer of Celestial Navigation*, New York: Cornell Maritime Press, 1944, 3rd edition.

E. G. R. Taylor, *The Haven-Finding Art*, New York: American Elsevier, 1971.

CHAPTER 15

SPACE TRAVEL

The newest application of astronomy, space travel is divisible into two types, the accomplished and the dream. The first, exemplified by voyages to the moon, was for centuries only in the minds of imaginative men and women. With the first human step on the moon, July 20, 1969, the idea came to fruition. In contrast to space travel within our own solar system, journeys to the stars and galaxies are thoughts of visionaries.

Early literature has plenty of references, such as the Tower of Babel in the Bible, to space travel. A voyage to the moon is described in *Vera Historia* by Lucian of Samosota, Syria (120–200) published in Greek about 160 A.D. and first translated into English in 1634. In a second moon story by the same author, the hero's transport to the moon is through the wings of a vulture and an eagle. His wings are taken from him after he also yearns to go to heaven. During the seventeenth century, astronomer Johannes Kepler (1571–1630) wrote his fantasy of a space voyage called *Somnium*. Bishop Francis Godwin (1561–1633) wrote *The Man In The Moon* and Bishop John Wilkins (1614–1672) was the author of *The Discovery of A World in the Moone*.

During the nineteenth century deeds rather than words characterized space travel. The rocket vehicle for space flight began to be developed.

The Chinese are alleged to have used powder rockets against the Mongols during the siege of Kaifeng in the thirteenth century but an improved solld-fuel rocket was developed in England at the start of the nineteenth century. (The Star Spangled Banner, written during the War of 1812 refers to the "rocket's red glare".) Rockets for space flight were studied by many Europeans including Konstontin Tsiolkowski (1857–1935) a Russian mathematics teacher, Hermann Ganswindt (1856–1934) a German law student, French Robert Esnault-Pelteric (1881–1957) and German Hermann Oberth (1894–   ). In 1923 the latter wrote "The Rocket Into Interplanetary Space". The American rocket expert Robert H. Goddard (1882–1945) wrote "A Means of Reaching Very High Altitudes" in 1919 and in March, 1926 successfully tested a liquid fuel rocket, probably the first such event.

The operation of rockets is described by Newton's third law of motion: For every action exists an equal and opposite reaction. Within a stationary closed expanded rubber balloon, for example, inside pressures on the surface are equal and opposed by a like amount of the rubber on the internal air. As soon as the neck of the balloon is opened, the air moves out and the pressures inside become unequal, the object moves in a random, zig-zag fashion until exhausted of air. A rocket is made of strong metal rather than rubber and its orifice allows for the controlled exit of the spent gases. Inside, the union of the fuel and supply of oxygen or similar oxidant are likewise controlled.

Rocket engines are rated by their thrust or force that can be generated to propel the device. Also important is the specific impulse, the thrust from each pound of propellant in one second of engine operation. Chemical fuels can be shown to have a theoretical limit of 400 for specific impulse. Two other significant rocket engine characteristics are exhaust velocity, a measure also of forward speed, and mass ratio, the comparison of the vehicle's mass at takeoff to the mass when all the propellant is burned.

Liquid propellants generally have higher energy values than do the solids. The liquids are also easier to control since their combustion can be cut off with the closing of a valve. Moreover, they are easier to cool since the propellant can be circulated through engine walls before injection so as to protect the walls from intense combustion temperatures. However, the liquids involve a more complex fuel injection system and the danger of handling them is greater.

Before human beings occupied the space vehicles, they were tested for safety and reliability. One of the leaders in this endeavor was John P. Stapp (1910–   ). He was in charge of the projects to determine the human consequences of extremely high speeds and accelerations, and was the guinea pig for many dangerous tests. Others underwent the trials of whirling in a centrifuge that could turn around at 175 miles per hour, equivalent to the stress of forty times the acceleration due to gravity. Experiments found that man in the prone rather than upright condition could withstand higher accelerations of the kind a rocket ship needs to escape the earth's gravitational pull.

By definition the escape velocity is that required to establish a parabolic orbit. Velocities greater than escape velocity result in hyperbolic orbits and lower ones give an elliptical orbit. The escape velocity from the earth is 36,700 feet per second and that from the moon is 7,800 feet per second.

Space travel beyond the moon necessitates unique launch velocities depending upon the destination. The minimum launch velocities, with transit times using chemical fuels are:

Mercury	44,000 feet/sec.	110 days
Venus	38,000 feet/sec.	150 days
Mars	38,000 feet/sec.	260 days
Jupiter	46,000 feet/sec.	2.7 years
Saturn	49,000 feet/sec.	6 years
Uranus	51,000 feet/sec.	16 years
Neptune	52,000 feet/sec.	31 years
Pluto	53,000 feet/sec.	46 years

The first successful manned voyage to the moon in 1969 was preceded by shorter jaunts into space closer to the earth. Those who tested not only collected scientific information but also pioneered in gauging the effects on human beings. The men and women are listed below.

First Twelve "Manned" Space Flights

Astronaut	Vehicle	Date	Time	Earth Orbits
Gagarin	Vostok 1	April 12, 1961	108 minutes	1
Shepard	Freedom 7	May 5, 1961	15 minutes	suborbital
Grissom	Liberty Bell	July 21, 1961	16 minutes	suborbital
Titov	Vostock II	August 6, 1961	25 hours 18 minutes	17
Glenn	Friendship 7	Feb. 20, 1962	4 hours 55 minutes	3
Carpenter	Aurora 7	May 24, 1962	4 hours 56 minutes	3
Nikoloyev	Vostock III	August 11, 1962	94 hours 25 minutes	64
Popovich	Vostock IV	August 12, 1962	70 hours 57 minutes	48
Schirra	Sigma 7	October 3, 1962	9 hours 13 minutes	6
Cooper	Faith 7	May 15, 1963	34 hours 20 minutes	22
Bykovsky	Vostock V	June 14, 1963	4 days — 2 hours 56 minutes	81
Miss Valentina Terechkova	Vostock VI	June 16, 1963	2 days — 22 hours 50 minutes	48

In the contest between the U.S.S.R. and the U.S. to reach the moon first, the Soviets at the start had the lead. From October 4, 1957, when they sent the first unmanned artificial satellite into space, to October 4, 1962, they sent 115,000 pounds of craft into orbit, disregarding secret ones. The U.S. record during the same time period was 102,000 pounds, excluding secret spacecraft. However, on July 20, 1969, American Niel Armstrong (1930—   ) became the first man to step on the moon and by the summer of 1971 four American landings had been made while the Soviets had sent only an unmanned satellite.

Three basic plans were considered by the Americans in going to the moon. Direct flight with direct lunar landing, earth-orbit rendezvous with direct lunar landing, and direct flight to a lunar orbit and descent to the moon's surface in a separate module. The last proved to the the quickest, cheapest and most reliable way for man to visit the moon.

Other solar system objects will be probed by instruments before man attempts to visit them. The same can be said for stars and galaxies but the problem is of an entirely different order. Even at the rate of 186,000 miles a second, that of light in a vacuum, the trip to the nearest star beyond the sun, Proxima Centauri, would take 4-1/3 years. Conditions do change at high speeds; according to relativity, time measures slow down. However, a device going even one-half the speed of light, is not yet available.

Space travel within our solar system, however, has brought advances not only in astronomy but also in biology, chemistry, meteorology, navigation, engineering and geology. For example, both the unmanned Earth Resources Technology Satellite (ERTS) and the manned Skylab satellite undertook remote sensing from space, and land use planning was immensely aided. ERTS-1 demonstrated its value in agricultural crop assessing, forest management, flood plain mapping, reservoir planning and electrical power plant locating. Other applications include predicting water runoff, uncovering mineral deposits, plotting extensions of geological faults and determining extent of eutrophication.

SELECTED REFERENCES

W. von Braun, *Space Frontier*, New York: Holt, Rinehart and Winston, 1971.
Arthur C. Clarke, *Man and Space*, New York: Time, Inc., 1964.

CHAPTER 16

UFOs AND ANCIENT ASTRONAUTS

After the end of World War II, a few men and women in the United States began to experience observations called "flying saucers." A small number of people insisted on interpreting the sightings as evidence for an extraterrestrial spaceship visiting the earth. When the number of people involved became larger, the United States Air Force began to collect, collate and investigate the phenomenon. Their Project Blue Book failed to satisfy the adherents to the visiting intelligence theme. A more exhaustive study at the University of Colorado under the direction of physicist Edward U. Condon, completed its report in 1968 and this, too, was similarly attacked.

Flying saucer is now a term of derision for misinterpreted sightings, for balloons, stars, planets, airplanes, birds, ball lightning and raindrops mistakenly viewed as guided by rational beings not living on earth. Unidentified flying objects, or UFO, is the more neutral phrase, without the connotation of an explanatory hypothesis.

UFOs have been seen in many countries and the records of past generations have been sifted to show their presence in ancient and medieval, as well as modern times. Of course, many of the sightings, then and now, can be explained away as familiar natural phenomena, but a small unexplainable percentage remains.

The uncritical can cite several supposed direct contacts between "spacemen" on UFOs and human beings. These include sexual intercourse with a "space woman" by a Brazilian farmer, conversation with the "space" people by a man who lived near Palomar Mountain in California, "probing" of two fisherman in Mississippi and the "examination" of a biracial married couple in New England. The validity of each of these varies but are as much established as the sighting of spacemen by a New Mexico policeman and the alleged cordial wave by a UFO occupant to a minister in New Guinea.

Comparable to the small number of "direct" contacts, are an equal amount of strange cessation of automobile power when near an UFO. Some people who have experienced the sightings reported also the inability of their cars to start or move.

No one has tested the alleged metallic nature of the UFOs by hurling a stone to detect a characteristic sound. Not a single piece of a vehicle exists to submit to laboratory analyses. Only imprints on soil and matted down grass are cited as effects of the UFOs parked on earth.

Unlike other astronomical categories such as planets, stars and galaxies, UFOs, even after subtracting those easily explained away, are apparently more than one variety of object. Thus, more than one hypothesis may be correct for them.

There is little likelihood that any of the UFOs are devices being tested by the military of this or any other country. The objects move with too great a speed and make sharp turns incapable of being done with present technology. Another suggestion that they are forms of life in the atmosphere, and perhaps the hydrosphere, would demand revision of the concept of life now generally accepted. Again, the high speed and turn characteristic are not known for life on earth.

Several other hypotheses can be laid aside as untenable. The sightings have been called projections of human fantasies because the UFOs resemble parts of the anatomy associated with sex. UFOs have been assigned to mass hysteria, modern fairy tales and signals for help to higher forces due to the stress and anxiety of modern life.

Assigning UFOs to an intelligence outside the earth is one idea that does not seem to go away. The concept describes the visions as spaceships propelled by beings alien to our planet. They could be either from our own solar system or another.

What are the chances of an intelligence in our solar system capable of mounting such space travel? On earth, we have only been able to travel to our moon since 1969, and we have not yet been able to send a manned vehicle elsewhere. Every other place in our solar system appears to be devoid of condition suitable for the specialized forms of life called higher and intelligent. Only Mars and Venus seem to be likely candidates as an abode for life, and indications of a civilization have not been detected on either planet.

The argument that unusual forms of life exist in our solar system so far not known to man, is not a feasible hypothesis. Not only is it contrary to the thesis describing life as DNA-based, but also the idea does not explain, describe or predict. It offers no help save a rationale for those who would explain away some UFO sightings.

Life is undoubtedly present in a multiple of solar systems, among the immense number in our observable region. (Although at the start of the twentieth century, American astronomers followed the lead of Henry Norris Russell who believed in the existence of only one solar system: ours.) The chance of life in another solar system, being able to travel to another and visit our planet, can be viewed from at least two perspectives.

The spread of the characteristic of life called intelligence or problem solving, undoubtedly follows the Gaussian distribution curve, the bell-shaped one. Every entity so far known, when large numbers are available, follows the identical pattern. Heights of men in a large city, income of American families, temperature of stars and the intelligence of life in the universe form a spread with a small number in a top category, a large middle group and a small number in a bottom slot. Life on planet earth as exemplified by man may be in any one of the three major classifications. If we are at the pinnacle group in problem solving then it is out of the question that some other life on another solar system has surpassed space travel capability. Despite the foolish activities assigned to mankind, we would not be surpassed in being able to travel away from our solar system, a feat we cannot do now. If man on earth be in the bottom section of the distribution, or even the middle one, a chance exists for more intelligent forms of life in the universe.

A second perspective based on more facts, helps delineate the position of man's intelligence in the distribution curve for all life. Ever since the middle of the twentieth century the universe has been thought to be derived from a gigantic expansion of primeval material. The big-bang hypothesis had until 1965 the evidence of the increasing red shifts of the galaxies; the spectral pattern indicated the expansion of the universe. In 1965 the radiation produced by the original cosmic explosion was detected, and the big-bang theory became more accepted.

If the universe did proceed from a common origin, then the galaxies and perhaps the billions of stars in each are approximately the same age. If each galaxy, and the stars in them, had the same general lines of development, then an advanced form of life is unlikely to occur in any particular solar system. However, the chance exists that some planets had more favorable conditions and civilizations beyond those on earth could have evolved. In our own galaxy, stars are of varying ages and presumably this is true in other galaxies. At least in our own island universe, the sun is about midway down the Russell-Hertzsprung diagram indicating a middle age. Older stars according to this pattern would be red giants and white dwarfs, and neither of these probably supports planetary systems. Stars like our sun or a bit older, are unlikely to have more advanced planetary civilizations.

Given the above factual and theoretical considerations, two alternatives for belief are available. First, is the untested conclusion that the small number of UFOs difficult to explain away are really space ships from another solar system. To hold such as the truth means belief in a life that has mastered high speeds and enormous gravitational attraction. The conception also raises a host of problems such as why the visitations to earth and why not during daylight hours. The number of scientific workers, astronomers, and others taken in by this doctrine, has never been counted; only an exceedingly tiny number have come forward publicly to give their support. A second conclusion views the Condon report as adequate explanation of the phenomena. An added proviso is that in all scientific endeavors a remnant of the inexplicable is present. Complete and total understanding of the natural world is an ideal for the scientific enterprise, has not been accomplished in very many realms, so why expect it for UFOs?

The belief that UFOs are spaceships has engendered another contention and "movement" that the earth was visited in the past by ancient astronauts who left artifacts of their stay. French writers first published the thought and it was taken up by other Europeans. A television program in the United States during 1972, spread the word so that clubs and organizations in support, similar to those for UFOs began to form. Another parallel is the presence of extremists in the ancient astronaut camp; some proclaim that God was a visitor from another solar system, with Jesus and other saints in all religions being ancient astronauts.

The flimsiest kind of evidence is presented to support the claim of ancient visitations. Formations in Peru, Egypt, and Easter Island are said to be the work of early superhuman explorers of the earth. The monuments cited can be explained as the work of earth people, and need not be assigned to extra-terrestrial beings. Thus, the Incas and their predecessors in South America made building on arable land a capital offense and so constructed housing in mountainous regions. They, as well as others who preceded us on earth, had great talents and cannot be put down as uncultured brutes a step away from the chimpanzee. The early Egyptians had their thing in building pyramids, just as the French under Louis XIV spent huge sums for Versailles. The monuments of Easter Island may seem as strange to us as our sports stadiums would be to the old Islanders. In short, our ancestors had building accomplishments along with such magnificent feats as the control of fire, the domestication of some plants and the discovery of the wheel.

Those who believe in ancient astronauts take up anomalies such as a young man found in the "wild" state in Western Europe, a gold bracelet found in coal beds millions of years old and a map showing the dry coast line of Greenland, under ice for thousands of years. Each of these facts has its own explanation and there is no need to involve an overall grand plan. The stratagem seemingly collates a number of inexplicable items, yet the explanation produces as many problems as are "solved." A Hans Castorp with tender soles and deficient native intelligence does not synchronize with a master race able to traverse huge, interstellar distances; a single gold bracelet should be near other artifacts of the same age; old maps are in the category as drawings on rock interpreted to be space helmets and radio antennae.

The ancient astronaut thesis has many bar-room tales presented as an adjunct truth. Men and women in a drunken stupor would find it easier to believe that stones with scratches found in Asia record the arrival of the astronauts several thousand years ago, that Oliver Cromwell in seventeenth century England fighting the forces of Charles II was guided by a hooded extra-terrestrial, that Baalbeck Monuments in Syria or the mountain-top strip in Peru were rocket platforms and landing fields.

All the adherents to the ancient astronaut idea present them with a technology and power of middle twentieth-century man on earth, and only the added advantage of space travel. The so-called early visitors have the laser beam, yet they built with stone rather than metal. They mated with a selected few earthlings rather than engage in genetic engineering.

Ancient astronauts and UFOs appeal to men and women who seek answers and explanation. This commendable drive, however, must be tempered with an understanding of evidence and the place of new ideas in the overall pattern of accepted knowledge. Some human cultures stressed omens such as the conditions of a bird's insides as a guide to action and belief. Others insisted that the truth was in whatever the forces of church or state promulgated. Only in relatively modern times has there arisen the necessity of belief based upon evidence. Even the heroes of western science, Copernicus, Galileo and Kepler, being a product of their times, did not fully practice the dictum that firm conclusions must be based upon evidence. It may well be that ancient astronauts existed but the evidence is lacking and therefore the concept cannot be a part of science.

An idea lacking evidence can be accepted in the frame work of human knowledge if it fits in and awaits the accumulation of data in its support. James Clerk Maxwell's concept during the nineteenth century was in this pattern for a couple of decades until Heinrich Hertz began to find the evidence; so indeed was the heliocentric conception at the time of Copernicus. But the ancient astronaut idea doesn't fit with other accepted material. If all the galaxies started to expand eighteen billion years ago, there is little likelihood of intelligences being formed at such great periods of time apart so that one is so far ahead of another.

There is one other possibility for truth for ancient astronauts. The field equations of general relativity, can help envision an astronaut leaving our universe, traveling along a path to a return almost anywhere in space-time. He could come back to the earth a billion years ago or a billion years in the future. But the difficulty again is the uniform age for our universe and thus small chances for life developing at vastly different rates.

If the ancient astronauts came out of other universes, then some general relativity concepts could help. A so-called Einstein-Rosen bridge from a black hole could be an entry port. But all of this is in the realm of science fiction, the same arena for ancient astronauts.

SELECTED REFERENCES

Carl Sagan and Thornton Page, Eds., *UFOs — A Scientific Debate*, Ithaca, N.Y., Cornell University Press, 1973.

The Condon Report, *Scientific Study of Unidentified Flying Objects*, New York, Bantam Books, 1969.

CHAPTER 17

ASTROLOGY

The characteristics of human beings are determined by heredity and environment. Scientists continually debate about the impact of each of these factors, but there is complete agreement about the total influence. Some of the genetic material passed from parent to child can be mapped and each site delineated for its effect. The place of environment is more broadly painted with such terms as "healthy surroundings" and "disadvantaged household."

The environment is thought to be the immediate area, but those who follow the precepts of astrology, point to certain groups of stars as the major factor, and discount heredity altogether. Astrologers proclaim human characteristics and events are determined solely through the stars where the sun apparently travels through the planets and the stars.

Astrological theory is not concerned with the sun, universally accepted as life giving. There is no doubt that without the sun all life on earth would not exist; that the amount of sunshine is an important agent in life support. All of this does not concern astrology. Nor is the doctrine involved with the effect of reflected light of the sun, principally from the moon. There are cultural legends claiming better crops when planted in the light of the full moon, or insanity (lunacy) from lunar rays (moon beams).

The chief thesis of astrology is that the zodiacal pattern at the time of birth indicates the future and character of the individual. This is applicable only to human beings and not to plants and animals.

Placing people apart from the remainder of life, is already at variance with the theory of organic evolution, the doctrine that life forms developed through time from unspecialized to specialized. What affects mankind should also be an influence for plants and animals, and vice versa. If radiation from certain stars does indeed interact with life on earth, a reason must be found why all below the great apes are exempted.

The stars in the zodiacal region are very distant from the earth. Radiation traveling at the enormous speed of 186,000 miles per second, needs years to come from the stars to our planet. A well-substantiated fact called the inverse square law, describes the attenuation of this radiation with distance; its intensity falls off inversely as the square of the distance. A radiation when two feet from its source is thus one-fourth as strong as when one foot from the origin. Radiation from the zodiacal stars is therefore exceedingly weak at the surface of the earth. Does it seem likely that such could determine the nature and future of a human embryo?

Astrologers may say that the influence of the stars is not causal, as suggested in the above paragraph. They do proclaim that there exists a correlation between an individual's life and the zodiacal sign. Men, women and children are said to be in the sign of Taurus, Virgo or Aquarius. Each of the twelve signs of the zodiac delineates a particular personality, with a closer analysis of exact time of birth giving individual characteristics.

Correlations are helpful in scientific work, with high, positive ones being held in esteem and explored for deeper significance and insight. Precipitation on earth is positively correlated with sunspots, and perhaps a better understanding of this relationship will one day be found.

The correlation of human birth with signs of the zodiac began to be established about 2000 years ago and astrology was born. At that time the section of the sky where the sun apparently travels was divided into twelve parts, with an arbitrary origin in Aries. At this junction the celestial equator (the earth's equator projected into the sky), intersects the line which is the sun's apparent path, the ecliptic. This intersection, the vernal equinox, shifts because of the earth's wobble called precession. Two thousand years ago, the first point of Aries was in one sky position. Today, because of precession, the zodiacal constellation Pisces is in its place. How can a correlation be maintained as correct under such circumstances?

Astrologers generally do not address themselves to the phenomenon of precession. Instead they cite the age and success of their work as well as the famous astronomers who practiced astrology. Truly, Johannes Kepler and Tycho Brahe cast horoscopes, but not a single modern astronomer is in the camp of the astrologers. Today, astronomers castigate astrology as a pseudo-science. Other critics contend that astrologers are not at all successful, with their predictions being vague and general. The long history of the venture, too, is not an adequate reason for supporting it because many social evils from war to slavery are ancient.

Astrology is generally enmeshed with mysticism, the supernatural, occult and irrational. In league with such anti-science forces, the subject is not open to the examination of its premises, makes no effort to be quantitative, has neither laboratory nor observatory for any on-going investigation and never bothers to check its results. As such, it is a subject in the same league with tea-leaf reading and crystal ball gazing.

SELECTED REFERENCES

Jack Lindsay, *Origins of Astrology*, New York, Barnes and Noble, 1971.
Louis MacNiece, *Astrology*, New York, Doubleday, 1964.

INDEX

2186 1